Knowledge Management and Organizational Learning

Volume 4

Series editors
Ettore Bolisani, Padova, Italy
Meliha Handzic, Sarajevo, Bosnia and Herzegovina

More information about this series at http://www.springer.com/series/11850

Ettore Bolisani • Constantin Bratianu

Emergent Knowledge Strategies

Strategic Thinking in Knowledge Management

Springer

Ettore Bolisani
Department of Management
and Engineering
University of Padua
Vicenza, Italy

Constantin Bratianu
Faculty of Business Administration
Bucharest University of Economic Studies
Bucharest, Romania

ISSN 2199-8663 ISSN 2199-8671 (electronic)
Knowledge Management and Organizational Learning
ISBN 978-3-319-60656-9 ISBN 978-3-319-60657-6 (eBook)
DOI 10.1007/978-3-319-60657-6

Library of Congress Control Number: 2017944182

Printed on acid-free paper

This Springer imprint is published by Springer Nature
The registered company is Springer International Publishing AG
The registered company address is: Gewerbestrasse 11, 6330 Cham, Switzerland

Foreword

The notion of knowledge is not new as well as its relevance for human beings as a means of survival and prosperity. A famous science-fiction film produced and directed by Stanley Kubrick *"2001: A Space Odyssey"* (1968) provides a good metaphor to point out the role and relevance of knowledge for society. The film starts with scenes of a tribe of ape-men, living in an African desert millions of years ago, which awakens to find a mysterious black monolith that has appeared before them. This monolith becomes their starting guide to learn how to use a bone as a weapon and, then, to get tools and methods to compete with their rivals and drive them away from the water hole, i.e., the source of their survival. The monolith is an exceptional metaphor to depict the role of knowledge for the human evolution. Knowledge is the key driving force of human life as well as of any transformational endeavors of our world. The capacity of creating, retaining, revising, applying, and projecting knowledge is at the basis of human being's survival, development, and progress. Human society has progressed by learning and embedding knowledge into whatever tangible and intangible entity is surrounding us. Every artificial object—as well as any notion of human inner or outer reality—is made of and it is the result of knowledge. This premise represents the underlying fundamental assumption of this book which offers a journey through nine chapters toward the analysis of knowledge, its strategic role, its use, and its strategic managerial deployment and exploitation to navigate the ever-increasing complexity of today's business landscape. It provides an important contribution to the management literature for both scholars and practitioners, by addressing two key conceptual pillars that advance the managerial understanding of how to manage and assess the knowledge-based drivers of organizational value creation mechanisms. Knowledge strategies and strategic thinking are proposed as two critical dimensions characterizing knowledge management.

In their authoritative book, Ettore Bolisani and Constantin Bratianu provide a clear outlook of the state of the art of the key conceptual pillars at the basis of the discipline of knowledge management. It represents a valuable resource both for scholars and for practitioners. Indeed, "what," "why," and "how" of knowledge management are thoroughly discussed. Although knowledge is not a new concept, it is fundamental to address and understand its meaning in the organizational and business context. Managers, in principle, are interested in knowledge not for the sake of knowledge but for the implications that managing knowledge can generate in terms of organizational performance improvements and value creation mechanisms. The understanding of "what" equals to clarify what is knowledge as an object whose relevance for managers and organizations is related to its role as a source and a resource of organizational wealth creation. Acknowledging the power of this concept, Bolisani and Bratianu point out that its definition is still very elusive. Different disciplines can concur to the definition of knowledge, ranging from social sciences to information theory, gathering the insights coming from philosophy, psychology, neurology, and sociology. All these disciplines can provide important perspectives of the notion of knowledge pointing out some specific features and traits, rather than disclosing its characteristics and building blocks components. From a managerial point of view, what matters is that knowledge is a strategic resource and source of company value creation and therefore is an organizational asset which acts as a fundamental strategic driver of competitiveness. In addition, since managers are interested in the practical implications of the deployment and exploitation of knowledge, they need conceptual tools to handle the notion of knowledge. From this point of view, the use of metaphors and of the metaphorical/analogical thinking, as proposed in this book, is of great relevance. The understanding of knowledge is the first step toward its use as a "strategic weapon" as proposed by Bolisani and Bratianu. This relates to the understanding of "why" knowledge management is necessary. Why organizations should increasingly be focusing their attention on managing their knowledge domains? The authors provide a clear explanation of the reasons that make knowledge a fundamental organizational value driver. Today's business context is characterized by increasing ambiguity, uncertainty, unpredictability, complexity, and turbulence, which make, overall, the business landscape in which organizations have to navigate more and more chaotic. In such a context, knowledge represents a critical success factor to survive and to drive growth. The acknowledgment of the knowledge-based nature of today's economy and of organizations points out that knowledge workers and knowledge processes are at the basis of organizations' competitiveness, and knowledge management is a necessary dimension connecting operations and strategy in order to translate knowledge into organizational performance, value outputs, and impacts. However, the ability of an organization to prosper is linked not only to its capacity of managing and developing its knowledge resources but most importantly to the identification of those cognitive resources that have strategic relevance for the future success of the organization and for this reason denote its knowledge assets. Indeed, knowledge assets contribute to the definition of the value of an organization from both a static and a dynamic perspective. From a static point of view, knowledge assets, such as patents, brand, culture, core competences, identity and image, and so on, stand for most of the market value of today's company. The notion of intellectual capital has been introduced in the management and economic literature to represent those assets explaining the difference between market value and book value of today's knowledge-intensive companies. On the other hand, and most importantly, knowledge assets define the roots of value creation dynamics. They are the value drivers to execute organizational processes that, in turn, explain organizational performance. The strategic management of knowledge assets, as discussed by Bolisani and Bratianu, is the way managers can attempt to understand the future. In particular, the authors propose the notion of knowledge strategy indicating that organizations should explicitly adopt a strategy to manage those knowledge assets affecting actual and future performance. This involves the definition of knowledge management initiatives as strategizing planned actions connecting business strategy and everyday operation management. The formulation and implementation of a strategy define the route that an organization's leadership undertakes to navigate the business landscape by coping with an unknown world and by continuously projecting and revising strategic objectives to be achieved. In order to operate, organizations need to develop their knowledge domains that in turn define organizational capabilities, skills, and competences. Knowledge strategies are aimed to maintain and develop organizational knowledge.

Although knowledge management is still not fully acknowledged as a mainstream managerial discipline with business schools dedicating courses on the subject, it represents a fundamental management and economic research area with fundamental implications to understand organizations, their working mechanisms and value creation dynamics, and the ways how organizations interact and shape the business landscape. This book, taking mainly a strategic viewpoint, reminds us that the real managerial relevance of knowledge and knowledge management is connected to the function of knowledge as an organizational value-driven source. Organizations are primarily in the business of knowledge and on their capacity of managing knowledge depends their future business sustainability and value creation capacity.

University of Basilicata
Potenza, Italy

Giovanni Schiuma

Introduction

For various reasons, this is an experimental book. First of all, it is experimental because it focuses on a topic—knowledge strategy—which is not new but still quite debated and controversial. Indeed, knowledge has long been accepted as a strategic asset to achieve and maintain competitive advantage. Drucker's (1969) anticipatory vision of a society, where traditional economic factors (i.e., land, labor, and capital) are complemented (or in some way replaced) by knowledge, has now become a reality: today is the age of the *knowledge economy* (Powell and Snellman 2004). There are new ways of considering labor, capital, and also technology. New models of the firm (Grant 1996; Senge 1990) become necessary. Novel challenges are posed to executives and decisionmakers. Intellectual capital and knowledge management (KM) gain their place in the practice of companies and in managerial research (Nonaka and Takeuchi 1995; Spender 2015; Davenport and Prusak 2000).

However, the notions themselves of both knowledge and strategy are quire unstable. Knowledge is an abstract concept, very powerful indeed, but without any reference to the tangible world and no clear definition so far. A popular view of knowledge, at least in the managerial disciplines, is that of "justified belief" (Nonaka and Takeuchi 1995). While this definition stems from earlier philosophical thinking, we should also consider that truth and its justification are, quite often, a matter of interpretation. So, there do exist different variations in the way we can see knowledge and represent it, depending on the particular context or situation.

As regards strategy, the term is one of the most frequently used in business, but, again, its definition is sort of dynamic. Apparently, it is clear that a strategy is important, especially because it resonates its military origins: we decide a vital goal and establish an appropriate way to achieve it. However, the possibility to do that—and, therefore, the usefulness itself of talking about a strategy—is influenced by some evident limitations. First, a strategy is intrinsically oriented toward an unknown future, and here the great impact of uncertainty has long been recognized (Mintsberg and Waters 1985). Second, those who formulate a strategy are not omniscient, and their capabilities are limited. So, a strategy is more a desired vision of the future rather than a rational formulation of a pathway to follow. As can be easily discovered in the managerial literature, the notion of strategy has changed over time, as a consequence of the challenges posed by the changing economic climate and, also, by the advancements in the theoretical reflections about the nature itself of strategic thinking.

In any case, it can be argued that the concept of knowledge is strictly intertwined with that of strategy: knowledge is not only an essential ingredient for formulating and even considering the idea of a strategy, but knowledge can also be the *object* of a strategy. In other words, there is increasing awareness that, for companies, it may be important to consider the new idea of a *knowledge strategy*.

Knowledge strategy is a concept that has started to become popular in the managerial literature only recently (Zack 1999) and mainly due to the upsurge of knowledge economy and the diffusion of knowledge management programs. While knowledge has always been a recognized ingredient of strategic formulation (as we mentioned, for deciding a strategy, we must *know* something), the notion of knowledge strategy means more. It suggests that a company should adopt a strategy to *manage its knowledge*. So, in addition to planning the production and delivery of products and services, deciding goals regarding profits and markets, and expressing objectives about competitive positioning, a knowledge strategy represents the effort to plan activities of KM and, more generally, to organize all resources and processes that, in a company, are devoted to developing knowledge and competences of people, boosting learning processes, and facilitating storage, sharing, and reuse.

Knowledge strategy is, however, a complex concept because it is a combination of two other complex concepts, i.e., knowledge and strategy. So, when it comes to defining appropriate approaches to formulating and implementing a knowledge strategy, everything becomes hard. This well explains why, at the beginning of this introduction, we declared that our book is experimental: it is about fluid concepts that still need to be stabilized in their boundaries and significance and clarified about their actual applicability.

But our book is experimental not only because of the topic but also because of the approach. We don't have an ultimate definition of knowledge strategy, nor we want to impose one. Rather, our purpose is to stimulate discussion and reflection in all those who may share the interest in this issue with us: researchers, practitioners, or students. We would like to discuss the state of the art of the debate on the topic, present and compare the various positions and viewpoints with an open-minded attitude, and, especially, we want to show how the notion of knowledge strategy is indeed complex, but its consideration and even application can provide food for thought to researchers and practitioners and can suggest new models and responses to the difficult challenges posed by our fast changing societies.

So, our real purpose is to stimulate the debate on what we consider a fascinating and fruitful concept and (possibly) to inspire others so that they can reach greater advancements in this field. Our modest ambition is simply that our book can be a honest contribution in this direction.

Although the whole volume centers on the concept of knowledge strategy, we decided that, to facilitate the reader, each chapter treats a specific point and has a separate list of references. Therefore, chapters can be read in sequence, or independently from one another: To help the reader in this, significant definitions and essential notions are often repeated in the various chapters, when they are important to understand the specific content. Chapter 1 introduces a discussion about the elusive notion of knowledge, which is, indeed, the starting point of our analysis. The intention is to show the most significant aspects of the dispute over the definition and the main conceptual barriers in that endeavor. Next, we show how knowledge has often been defined by using metaphors, and this approach has important implications also in the practical interpretation and use.

Chapter 2 focuses on KM, which rapidly became a sort of necessity in the postindustrial society and in the knowledge economy. However, we adopt a heterodox approach: instead of starting from defining KM, describing its functions, and, in the end, proposing a prescriptive framework, the chapter depicts the broad picture of the changes in the structure of the economy, where tangible resources are increasingly replaced by intangible resources as the main production factors. The new attributes of knowledge workers and knowledge processes that become vital in knowledge-based organizations are then discussed. So, since knowledge and its functions constitute strategic resources, knowledge management bridges the gap between operational management and strategic management for knowledge-based companies.

The purpose of Chap. 3 is to explore the uneasy connection between knowledge and strategy. Knowledge is increasingly considered a fundamental strategic weapon for value creation, especially due to the increasing turbulence of the business environment. This soon brings us to the idea of a strategy as a means of realizing it. A discussion about the notion of strategy is then provided, and we will show how this notion has evolved over time as a result of a new equilibrium of forces between internal and external business environment. Here, knowledge has always been a central element in all the various perspectives on strategy and strategic thinking that can be found in the literature and in the managerial practice of companies.

As we said before, strategies imply a vision of the future: they are built for future actions, and so understanding the nature and the content of future becomes important. In Chap. 4, we discuss our perception of time and we show how we understand the concept of future within the framework of time, complexity, and uncertainty. Human mind developed, during its historical existence, a series of metaphors able to suggest new semantic dimensions of time and its role in structuring the future. Due to the complexity of the future, it is also important to explain the way we correlate variables describing events and phenomena that we would like to predict for future decisions. This also means to explore the paradigms of linearity and nonlinearity or, in other words, the different views we can adopt to connect past and future and ground our forecasts. Finally, we present the semantic dynamics of uncertainty: uncertainty has an unavoidable role in defining probable futures, and the literature of business management has long considered uncertainty in the definition of strategic objectives and in the design of strategies able to achieve them.

The second part of the book addresses its core topic directly. Chapter 5 treats the notion of knowledge strategy. In addition to the classic elements of a strategy in a company, a knowledge strategy represents the effort to plan activities of knowledge management and, more generally, all resources and processes that are devoted to developing knowledge and competences of people, boosting learning processes, and facilitating storage, sharing, and reuse. As the chapter shows, the definition of knowledge strategy is difficult, and it is still necessary to clarify its contents and boundaries. In addition, there is the need to explain if a knowledge strategy is just a part—or a derivation—of classic strategic formulation of companies, or if it must have a special and distinct place that also deserves specific approaches and methods.

Considering knowledge strategies also leads to another point: how can a knowledge strategy be planned and formulated? What approach can be more appropriate? Chapter 6 focuses on *strategy formulation*, i.e., the process of formulation of a strategy for a company. As we will have learned in the previous parts of the book, the intrinsic presence of sources of uncertainty and turbulence finally affects the way strategy formulation can be approached consistently. And so, the introduction of the notion of knowledge strategy poses additional challenges. Can it be of some help for strategists, or does it just add new complications to strategy formulation? How can a knowledge strategy be formulated? Is its formulation related to the usual processes of strategy formulation, or is it somewhat special? And how is the formulation of a knowledge strategy related to that of a company's general strategy? All these questions are treated in this chapter. Particularly, the so-called deliberate or rational approach to knowledge strategy (and, more generally, to strategic planning) is discussed, as well as its limitation. The idea of knowledge strategy as an "emergent strategy" in companies (which also explains the title of the book) is then presented. Finally, we propose a way to integrate both the approaches to strategy and knowledge strategy (i.e. deliberate/rational and emergent), for better fitting the challenges of a dynamic, turbulent and uncertain environment.

To formulate a knowledge strategy, it may be useful to adopt some standard references. In classic textbooks of strategic management, different possible strategic options are named and classified in general terms, so that they can serve as a basic starting point for strategic design and formulation in real companies. This is exactly the same approach that is proposed in Chap. 7, which focuses on *generic knowledge strategies*. In substance, the chapter proposes a review of the literature, where a number of standard options have been analyzed, classified, and named appropriately. So, this can help strategists to become more aware of the role of knowledge strategies in companies and to formulate their own strategy based on variations and adaptations of one or more generic knowledge strategies. The generic knowledge strategies presented in this chapter are the following: exploitation strategies, acquisition strategies, sharing strategies, and exploration or knowledge creation strategies. The readers who are more familiar with knowledge management will immediately recognize these classic terms, which are becoming quite popular in the field.

Finally, Chap. 8 treats an apparently disjointed topic—that or measurement—which is, however, integral part of strategic management: it is generally assumed that we need a method to measure the strategic performances that define our goals, to control the implementation of a strategy, and to assess its final success. So, when it comes to knowledge strategies, we may say that we need a way to measure strategic performances related to knowledge. This means we need some measurement system that can be applied to knowledge and knowledge management. But if and how it is possible to measure knowledge is still questionable. Many techniques have been proposed, but they are far from becoming an established practice, and even a *rationale* to treat the problem still lacks. The chapter gives some order to this much debated issue and analyzes the theoretical and methodological soundness of the various measurement techniques.

In concluding this presentation, a few words of acknowledgement of all those that made this book project possible. First of all, Springer's team of Editors and Editorial Assistants, who have supported it with their proactive help. Second, we gratefully thank many of our co-members of the International Association for Knowledge Management for the feedbacks and reviews they provided in various moments of this long and complex project. Particularly, we would like to thank our colleagues and friends Meliha Handzic, Coeditor of the Book Series, for her warm encouragement and support, and Giovanni Schiuma, who kindly wrote the foreword. Finally, we can't forget our beloved Families, who—as usual—didn't miss to provide their support and encouragement even in the hardest times of work.

References

Davenport, T. H., & Prusak, L. (2000). *Working knowledge: How organizations manage what they know*. Boston, MA: Harvard Business School Press.

Drucker, P. F. (1969). *The age of discontinuity*. Piscataway, NJ: Transactions Publishing.

Grant, R. M. (1996). Toward a knowledge-based theory of the firm. *Strategic Management Journal, 17*(S2), 109–122.

Mintzberg, H., & Waters, J. A. (1985). Of strategies, deliberate and emergent. *Strategic Management Journal, 6*(3), 257–272.

Nonaka, I., & Takeuchi, H. (1995). *The knowledge-creating company: How Japanese companies create the dynamics of innovation*. Oxford: Oxford University Press.

Powell, W. W., & Snellman, K. (2004). The knowledge economy. *Annual Review of Sociology, 30*, 199–220.

Senge, P. (1990). *The fifth discipline: The art and science of the learning organization*. New York: Currency Doubleday.

Spender, J. C. (2015). Knowledge management: Origins, history, and development. In E. Bolisani & M. Handic (Eds.), *Advances in knowledge management* (pp. 3–23). Berlin: Springer International Publishing.

Zack, M. (1999). Developing a knowledge strategy. *California Management Review, 41*(3), 125–145.

Contents

1 The Elusive Definition of Knowledge 1
1.1 Searching for Knowledge Definition 1
1.1.1 Knowledge Nature 1
1.1.2 Knowledge Definition 3
1.1.3 Three Kinds of Knowledge 5
1.2 Knowledge Metaphors 7
1.2.1 Metaphorical Thinking 7
1.2.2 Knowledge as Objects 9
1.2.3 Knowledge Nuggets 11
1.2.4 The Iceberg Metaphor 12
1.2.5 Knowledge Flows 13
1.3 The Energy Metaphor 15
1.3.1 Knowledge as Energy 15
1.3.2 The Field of Rational Knowledge 16
1.3.3 The Field of Emotional Knowledge 17
1.3.4 The Field of Spiritual Knowledge 18
1.4 Conclusion 19
References 21

2 The Emergence of Knowledge Management 23
2.1 The Knowledge Economy 23
2.1.1 The Fourth Discontinuity 23
2.1.2 Key Features of the Knowledge Economy 25
2.2 The Knowledge Organization 27
2.2.1 The Economic Theory of the Firm 27
2.2.2 The Knowledge-Based Theory of the Firm 30
2.2.3 The Learning Organization 33
2.3 Knowledge Management 35
2.3.1 Knowledge Workers 35
2.3.2 Knowledge Processes 37
2.3.3 Knowledge Management as an Emergent Process 40
2.4 Conclusion 43
References 44

3 Knowledge as a Strategic Weapon 49
3.1 Strategic Knowledge and Value Creation 49
3.1.1 Knowledge as a Strategic Resource and Capability 49
3.1.2 Knowledge Value Chain 53
3.1.3 Strategic Knowledge and Intellectual Capital 55
3.2 Strategy as a Dynamic Notion 57
3.2.1 The Military Origin of the Concept 57
3.2.2 The Extension of Strategic Thinking to Business 59
3.3 The Knowledge-Based View of Strategy 61
3.3.1 Classical Streams in Economics 61
3.3.2 The Age of Strategic Planning 62
3.3.3 Planning Strikes Back: Competitive Strategies and Resource-Based Views 64
3.3.4 Strategy as Practice 66
3.4 Lessons from a Long History 67
3.5 Conclusion 68
References 69

4 Understanding the Future for Strategy Formulation 73
4.1 Time Perception 73
4.1.1 Inertial Thinking 73
4.1.2 Dynamic Thinking 75
4.1.3 Entropic Thinking 76
4.1.4 The Flying Time 78
4.2 Dealing with Complexity 81
4.2.1 Linear Thinking 81
4.2.2 Nonlinear Thinking 83
4.2.3 Systems Thinking 85
4.3 From a Single Certain Future to Multiple Probable Futures 86
4.3.1 Deterministic Thinking 86
4.3.2 Probabilistic Thinking 87
4.3.3 Probable Futures 89
4.3.4 Future as Chaos 90
4.4 Conclusion 92
References 93

5 Knowledge Strategies 97
5.1 An Important Notion but a Difficult Definition 97
5.2 Knowledge Strategy or KM Strategy? 104
5.3 Knowledge Strategy and Company Strategy: Which Comes First? 106
5.4 A Tentative Summary 112
References 113

6 Knowledge and Strategy Formulation in a Turbulent World 117
6.1 Strategists Under the Sword of Damocles 117
6.1.1 How to Formulate a Strategy? From Strategic Planning to Strategic Management, and Beyond 119
6.1.2 The Appeal of Deliberate and Rational Strategy Formulation 121
6.1.3 Emergent Strategy Formulation 123
6.1.4 Multiple and Parallel Formulations 125
6.1.5 Planning Levels 127
6.1.6 Attempt of a Synthesis 128
6.2 Knowledge Strategy Formulation 128
6.3 Knowledge and Strategy Formulation 131
6.3.1 Paths and Patterns in Strategy Formulation 132
6.3.2 Analysis 136
6.4 An Integrated Model: Formulating Strategies in the Learning Organization 139
6.5 A Summary 143
References 143

7 Generic Knowledge Strategies 147
7.1 Thinking Perspectives 147
7.1.1 Generic Business Perspectives 147
7.1.2 The Known-Unknown Matrix 149
7.1.3 Organizational Knowledge Dynamics 151
7.2 Knowledge Exploitation Strategy 152
7.2.1 Organizational Ambidexterity 152
7.2.2 Knowledge Codification 153
7.2.3 Knowledge Mapping 157
7.3 Knowledge Acquisition Strategy 159
7.3.1 Knowledge Acquisition 159
7.3.2 Knowledge Capturing 160
7.3.3 Knowledge Retention 162
7.4 Knowledge Sharing Strategy 163
7.4.1 Knowledge Sharing 163
7.4.2 Communities of Practice 164
7.5 Knowledge Exploration Strategy 167
7.5.1 Knowledge Creation 167
7.5.2 Knowledge Co-creation 168
7.6 Conclusion 169
References 170

8 Strategic Performance and Knowledge Measurement 175
8.1 Measurement: An Essential Ingredient of Strategies? 175
8.2 Knowledge Measurement for Knowledge Strategies 179
8.3 Knowledge Measurement in the Literature 180

8.4 Existing Measurement Methods 183
8.5 Knowledge Measurement and Theories of Measurement 190
8.6 Perspectives 194
References 196

9 Beyond Conclusion 199

Author Biography

Ettore Bolisani is Associate Professor at the University of Padova. Previously, he was the recipient of the prestigious European Union "Marie Curie" research fellowship at Manchester University and a researcher at the Universities of Trieste and Padova. He was also a visiting researcher at Coventry University, visiting lecturer at Kaunas Technological University, and the chair of the 2009 European Conference on Knowledge Management (2009). He is the cofounder and first President of the International Association for Knowledge Management (http://www.iakm.net) and is the Series coeditor (with Meliha Handzic) of the IAKM Book Series on *Knowledge Management and Organizational Learning* published by Springer. His research focuses on innovation management and knowledge management.

Constantin Bratianu is Professor of Strategic Management and Knowledge Management at the Bucharest University of Economic Studies, Romania. He is the founding Director of the Research Centre for Intellectual Capital and Entrepreneurship and former Director of the UNESCO Department for Business Administration. Previously, he was a visiting professor at Universities in the USA, Japan, Austria, Hungary, and Egypt. He is the coeditor of the international journal *Management Dynamics in the Knowledge Economy* and Chair Program of the 2016 European Conference on Management, Leadership and Governance, which was held in Bucharest. His main academic interests span across the fields of knowledge dynamics, knowledge management, intellectual capital, learning organizations, and strategic management.

The Elusive Definition of Knowledge 1

Knowledge is an abstract concept without any reference to the tangible world. It is a very powerful concept, yet it has no clear definition so far. From the Greek philosophers up to present experts in knowledge management, people tried to define knowledge but the results are still very fuzzy. This chapter has the intention of showing the most significant aspects of the dispute over the definition of knowledge, and the main conceptual barriers in that endeavor. In the first part of the chapter we discuss about the knowledge nature and the attempts made in epistemology to define knowledge. The well-known definition that *knowledge is justified true belief* is shown to have the limitations given by the justification condition and the truth nature. In the second part, we consider the metaphorical approach to knowledge explanation and we present the main metaphors used for knowledge in the managerial literature: knowledge as objects, knowledge nuggets, knowledge as an iceberg, and knowledge as stocks and flows. In the last part, we introduce a new paradigm of metaphorical thinking based on the knowledge energy. This metaphor opens new opportunities for understanding knowledge as a multi-field paradigm composed of the rational, emotional, and spiritual knowledge fields.

1.1 Searching for Knowledge Definition

1.1.1 Knowledge Nature

Knowing is one of the most specific human processes and *knowledge* is its result. That means that knowing and knowledge have been subjects of human inquiry from the ancient times. Some great philosophers, starting with Plato and Aristotle, developed *Epistemology* as a theory of knowledge, trying to answer to the fundamental question: What is knowledge? There were many answers and many arguments used in supporting their theories, but none of those has been accepted so far as being fully satisfactory. Defining knowledge and explaining its nature proved to be elusive and without a convincing and universally accepted result (Neta

E. Bolisani, C. Bratianu, *Emergent Knowledge Strategies*, Knowledge Management and Organizational Learning 4, DOI 10.1007/978-3-319-60657-6_1

and Pritchard 2009; Russell 1972). Most of the theories have been integrated into two major perspectives: rationalism and empiricism. Simplifying, we may say that both theories accept that knowledge is *a justified true belief*, but they depart in showing the ways through which one can find the truth or justifying the true belief.

Rationalism, for which Plato is a pioneering philosopher, argues that knowledge is a result of a reasoning process and that our sensory experience plays no role. Knowledge can be obtained only from rational reasoning grounded in axioms, like in mathematics, and it should be distinguished from opinion which is a product of our senses. In his theory about ideas, Plato makes a difference between a "cat" which represents a particular object in the real world and the concept of "cat" coming from the eternal world of *cattyness*. While the real "cat" is born and sometimes will die, the concept of "cat" remains in the eternal world of ideas. Knowledge belongs to that eternal world. Explaining Plato's framework of knowledge, Bertrand Russell (1972, p. 152) shows that "We perceive hard and soft through touch, but it is the mind that judges that they exist and that they are contraries. Only the mind can reach existence, and we cannot reach truth if we do not reach existence". We cannot know the real world through senses alone since they can mislead us. In conclusion, "knowledge consists in reflection, not in impressions, and perception is not knowledge" (Russell 1972, p. 153). We may agree with Plato when discussing about mathematics and mathematical propositions. To understand that $y = a + bx$ reflects a correlation between two variables we don't need any sensory perception. We need only a reasoning process with abstract symbols. But that is just a particular domain of science and cannot be generalized over the whole human existence.

René Descartes made rationalism the basis of modern philosophy by integrating many new scientific discoveries in his conceptual universe. He founded the famous method of doubting everything and searching for certainty: "I can do nothing else, until I have learned for certain that there is nothing in the world that is certain" (Descartes 1997, p. 139). By analyzing comparatively his thoughts coming from the mind and the information coming from the sensory system, Descartes reached the conclusion that thought is the only attribute that belongs to him and cannot be detached from him: "What of thinking? I find here that thought is an attribute that belongs to me; it alone cannot be separated from me. I am, I exist, that is certain" (Descartes 1997, p. 141). That means that the only evidence of our existence is the fact that we think and through thinking we acquire knowledge. In his famous formulation "*Cogito, ergo sum!*", mind and body are like two different worlds, and while bodily sensations fail the reliability test, thinking proves to be the unique characteristic that is reliable and certain. Finally, he remarks: "I am, however, a real thing and really exist; but what thing? I have answered: a thing which thinks" (Descartes 1997, p. 142). This dualism of mind and body had a great impact on science, philosophy and education in Europe, and later on in America. Even today, many authors consider knowledge to be rational and based on solely mental processes.

Empiricism emerged as an opposable perspective to rationalism. Aristotle, a former student of Plato, considered that ideas and forms cannot be separated from physical objects and sensory information. Knowledge is not created a priori and is not innate in a deterministic form. It is created through our sensory interface with the real world, and it is processed finally by our mind. John Locke continued that approach emphasizing that objects do exist in the outer world and that our sensory perception is the most important source of our knowledge. Many contemporary philosophers tried to bridge the gap between rationalism and empiricism by generating conceptual frameworks based on different syntheses between them.

In sharp contrast with the Cartesian dualism of mind and body, the Japanese intellectual tradition based on Buddhism and Confucianism created an integrated perspective of mind and body with three overarching premises (Nonaka and Takeuchi 1995, p. 27): "(1) oneness of humanity and nature; (2) oneness of body and mind; and (3) oneness of self and other. These traits have formed the foundation of the Japanese view toward knowledge as well as the Japanese approach toward management practices". That means that knowledge is rooted in the sensory system and only in its final processing stage is open to abstract considerations. Their relation with the real world is through their senses and they don't need to make appeal to any eternal or metaphysical world in order to understand the nature of knowledge. Mind and body are not two distinct realities but an integrated one which creates the whole personality of people. "For the Japanese, knowledge means wisdom that is acquired from the perspective of the entire personality. This orientation has proved a basis for valuing personal and physical experience over indirect, intellectual abstraction" (Nonaka and Takeuchi 1995, p. 29). This integrated view can be seen in the *samurai* education, where internal meditation was used together with physical training, and in the knowledge management practices developed within Japanese companies where the focus is on tacit knowledge which reflects the best people's direct experience. It is interesting to see how Miyamoto Musashi, the legendary Japanese martial artist, emphasizes in his famous *Book of five rings* the importance of learning with the whole body the correct motion during a fight (Kaufman 1994, p. 31): "Proper movement of the body depends entirely on the manner in which you carry yourself. The feet carry the body and the body directs the feet. Tread firmly with the heel touching the ground first and then roll forward to the ball of your foot. Practice this until you appear to move without motion".

1.1.2 Knowledge Definition

As mentioned before, a frequently adopted definition of knowledge is that of "justified true belief" (Nonaka and Takeuchi 1995, p. 87). That definition incorporates three basic conditions, fact for which some authors call it the *tripartite account of knowledge*. These conditions are the following (Neta and Pritchard 2009).

- *The truth condition*. It requires that if one knows a proposition then that proposition must be true. If the proposition is not true, then that person does not know what he claims to know. The truth condition makes the difference between opinion and knowledge.

- *The belief condition.* That condition demands that if one knows a proposition then he believes that proposition.
- *The justification condition.* That condition requires a practical way of justifying that the belief one has is true.

Putting together these conditions for knowing, one may conclude that "the necessary and sufficient conditions for knowing that something is the case are first that what one is said to know be true, secondly that one be sure of it, and thirdly that one should have the right to be sure" (Ayer 2009, p. 13). The right to be sure can be earned in different ways which are culturally and contextual dependent. These conditions are usually synthesized in a logical format. Considering S to be the subject or the knower, P to be the proposition the subject is supposed to know, one may write (Gettier 2009, p. 14): S knows P if:

- P is true,
- S believes P, and
- S is justified in believing P.

However, Gettier (2009) constructed some counter-examples to demonstrate that this formulation does not constitute a sufficient condition for the subject S to know P since justification might not be reliable. A person may be completely justified in believing something (i.e. P) which can be false. In literature, this case is known as the "Gettier problem" with respect to justification. Lehrer (2009) introduces a fourth condition to solve that problem, but it is too abstract to discuss it here. It is much more appealing to discuss how Nonaka and Takeuchi (1995) consider the justification problem in practice, which means in a company.

In their famous theory of organizational knowledge creation, Ikujiro Nonaka and Hirotaka Takeuchi (1995) adopted, for knowledge, the classical definition formulated by Plato that "knowledge is justified true belief". However, there is a significant difference in interpreting that definition. While the Western epistemology focuses on *truthfulness* as being the main characteristic of knowledge, Nonaka and Takeuchi (1995, p. 58) focus on *justified belief* arguing that: "While traditional epistemology emphasizes absolute, static, and nonhuman nature of knowledge, typically expressed in propositions and formal logic, we consider knowledge as a dynamic human process of justifying personal belief toward the truth". In other words, the authors change the philosophical discussion into a managerial practice and consider that the best way of justification is against the social context where new knowledge is created and shared, which means the organizational context. However, by doing this switch the authors show that, in practice, the emergence of new knowledge should be evaluated with a usefulness metric and not with a logical one: "Justification involves the process of determining if the newly created concepts are truly worthwhile for organization and society" (Nonaka and Takeuchi 1995, p. 86). They go further giving as practical justification criteria like cost, profit margin, and degree to which a product can contribute to the company's economic performance. However, by means of this switch, they changed the very nature of justification from a logical construct to an economic one, implemented by managers. Top managers would ask for a concordance with the strategic vision of the company, while the middle managers would be looking for some practical requirements. In conclusion, the approach of Nonaka and Takeuchi clearly changed the nature of the problem and offered solutions for the practical organizational context instead of solving the original truthfulness problem formulated by Plato and refined by the Western epistemology. It is like Alexander the Great who not being able to unfold the famous Gordian knot cut it with his sword and changed the history of the world.

We see that truth and its justification is mostly a matter of interpretation, and although the epistemological approach looks like a precise and logical formulation the final definition of knowledge may be just an illusion. The truth is far away and can be distorted by the justification attempt due to misunderstanding of the organizational context. Metaphorically, we may think of the *Fata Morgana* phenomenon. *Fata Morgana* is a mirage that appears on land or at sea, in deserts or in polar regions. It is an optical phenomenon resulting from the passage of the light rays through layers of air of different temperatures. In essence, it manifests as inverted floating images right above the horizon. Metaphorically, defining knowledge may result in such a mirage since considering the framework of epistemology we may already have different layers of relative truths.

The definition of knowledge remains a problem, at least in the managerial sense, since knowledge, becoming a strategic organizational resource, needs to be defined as an operational concept adequate for a business environment and not as an abstract one for a transcendental world of ideas. Knowledge definition is elusive since premises for initial conditions have been formulated on pure rationalistic grounds and a Cartesian perspective on human nature. In the following sections of this chapter we will change the conceptual paradigm of Greek philosophers with the new paradigm of cognitive sciences and will continue our journey to finding a better definition for knowledge.

1.1.3 Three Kinds of Knowledge

Adopting an integrated view on the nature of knowledge, some authors (Dombrowski et al. 2013) explain that there are three kinds of knowledge: (a) experiential knowledge; (b) skills; and (c) knowledge claims. They are interconnected, but have some specific features of their own.

Experiential knowledge is what we get from the direct connection with the environment, through our sensory system, and then it is processed by the brain. For instance, if we want to know what snow is then we must go where there is snow and touch it, smell it, taste it and play with it. We cannot get that knowledge only from books or seeing some movies with people enjoying winter sports in beautiful mountain areas. People living in geographical zones where there is never snow have real difficulties knowing what snow is. They lack the experiential knowledge about snow. Experiential knowledge is personal since it can be acquired only through direct interface of our sensory system and then processed by our brain. It is essentially based on perception and reflection. Several people having together the same experience may acquire different experiential knowledge since reflecting upon a living experience means actually integrating it in some previous similar experiences and knowledge structures, if they do exist. "Things are not always as they appear to be and our own perspectives influence our interpretations. Still, watching out for errors in thinking can improve tremendously the quality of our reflections on our experiences" (Dombrowski et al. 2013, p. 38). As we will show later, experiential knowledge can be seen as created by a powerful interaction between emotional, rational and spiritual knowledge since it is a result of the whole body and mind active participation (Bratianu 2015).

Skills means knowledge about *how to do* something (know-how). It is based on experiential knowledge but it is a well-structured and action oriented knowledge we get by performing repeatedly a certain task and learning by doing it. This is the way of learning swimming, biking, skiing, playing piano or doing many other similar activities. It is like learning unconsciously to perform a certain procedure or to follow a given algorithm. We don't learn swimming by reading in a book about fluid mechanics and objects floating. We have to learn by doing it with the whole body and reflecting upon it to improve coordination between breathing and moving our arms. *Know-how* knowledge is often called *procedural knowledge* since it is about performing a task in concordance with a given procedure or algorithm. We discussed about some skills associated to physical activities but they can be developed for any kind of task or activities, including thinking processes. For instance, thinking skills are extremely important for knowledge workers and decision makers. One of the most important skill in designing strategies is *intuition.* According to Klein (2003, p. 36), "The key to using intuition effectively is experience—more specifically, *meaningful* experience that allows us to recognize patterns and build mental models. Thus, the way to improve your intuitive skills is to strengthen your experience base. The most meaningful type of experience, naturally, is real-life experience".

Knowledge claims are what we know, or we think we know. We don't know how much we know since knowledge means both explicit knowledge and tacit knowledge, which means experience existing in our unconscious zone and manifesting especially as intuition. Explicit knowledge is something we learn in schools and reading books, or just listening to some professors or conference speakers. Knowledge claim is what we frame in an explicit way by using a natural or symbolic language. Thus, language is an essential component of the transforming our emotional and spiritual experience into rational or explicit knowledge. With explicit knowledge we are entering the zone of exchange between personal and shared knowledge. "Because ideas are stated in language, they can be examined and discussed, questioned, evaluated, refuted, or published and passed on. Knowledge claims enable us to learn from each other and built our shared knowledge" (Dombrowski et al. 2013, p. 44).

1.2 Knowledge Metaphors

1.2.1 Metaphorical Thinking

Cognitive scientists discovered that thinking is a conceptual process which is primarily metaphoric. That means that metaphors represent much more than just linguistic expressions. They are involved in our thinking process, helping us to understand new concepts and ideas. Steven Pinker, a famous cognitive scientist and professor at the Department of Psychology at Harvard University, explains that: "Conceptual metaphors point to an obvious way in which people could learn to reason about new, abstract concepts. They would notice, or have pointed out to them, a parallel between a physical realm they already understand and a conceptual realm they don't yet understand" (Pinker 2008, p. 241).

Fundamentally, metaphors are embodied in our experience and through a progressive abstraction process they lead to new meanings for less known objects or concepts. As underlined by Lakoff and Johnson (1999) in their captivating book *Philosophy in the flesh. The embodied mind and its challenges to western thought*, any complex metaphor can be decomposed into primary metaphors, and "each primary metaphor is embodied in three ways: (1) It is embodied through bodily experience in the world, which pairs sensorimotor experience with subjective experience. (2) The source-domain logic arises from the inferential structure of the sensorimotor system. An (3) it is instantiated neutrally in the synaptic weights associated with neutral connections" (Lakoff and Johnson 1999, p. 73).

Metaphors are similar to analogies which create comparisons between a known object or concept and a less known one. They allow us to map one experience in terms of another experience, making it possible to understand complex and new situations in terms of what we already know. A metaphor is composed of two different semantic domains: (a) a source domain where we describe the known object or concept with its structural and functional attributes, and (b) a target domain where we place the less known object or concept. Metaphorical thinking means to analyze the attributes and relationships from the source domain and to compare them with the situation from the target domain trying to identify which of these elements can be transferred from the source domain into the target domain. Theoretically, we perform a structural mapping of the known attributes and relationships from the source domain onto the target domain (see Fig. 1.1).

As a result of this process, the less known object or concept receives new semantic attributes which lead to its better understanding. As Lackoff (1990) suggested, metaphors can create meaning and enlarge the semantic horizon of the less known object or concept. That means that, in a metaphorical process, a conceptual system is projected from one domain to another, which is usually more abstract. It is a progressive abstraction effort, which will be clearly demonstrated in the case of knowledge metaphors (Gentner et al. 2001). However, not all structural and functional attributes from the known semantic domain can be transferred into the less known semantic domain which means that we discuss about a selective mapping based on some sound hypotheses and principles. For instance,

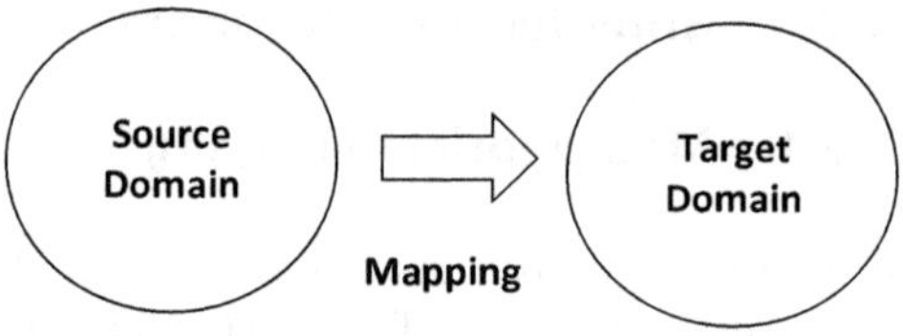

Fig. 1.1 The structure of a conceptual metaphor

in the well-known metaphor *Time is money*, the source domain contains the semantic field of the concept *money*, and the target domain contains the semantic field of the concept *time*. In this metaphor, *money* represents a tangible object with some physical or structural attributes and some functional or intangible ones. *Time* represents an intangible object only with intangible structural and functional attributes. Thus, the metaphor cannot map the physical attributes of money onto the target domain, but it can map the functional intangible attributes like *spending* and *saving*. For instance: *I saved one hour by driving the car on a different route.*

The process of structural mapping from the source domain onto the target domain is unidirectional and asymmetric. It is *unidirectional* since mapping is done only in one way according to our purpose to enlarge the semantic field of the less known concept. It is *asymmetric* since the target domain has a deficit of semantic attributes by comparison with the source domain. By means of structural mapping, the degree of asymmetry is decreased and the target domain is enriched with new semantic attributes. We will illustrate this phenomenon in the following sections with some significant knowledge metaphors. Knowledge is an abstract concept with no physical counterpart. Defining knowledge from pure theoretical point of view proved to be difficult and fuzzy, especially when interpreting the justification condition. Metaphorical thinking opens a new way of understanding and defining knowledge by placing it in the target domain and searching for meaningful tangible or intangibles entities placed in the source domains. But that means that there is an endless series of objects and concepts which can be used in the source domain, and that knowledge definition depends on the metaphor used for its explanation. As Andriessen and Boom show, "Knowledge is not a concept that has a clearly delineated structure. Whatever structure it has it gets through metaphor. Different people from different cultures use different metaphors to conceptualize knowledge. They may be using the same word; however, this word can refer to totally different understandings of the concept of knowledge" (Andriessen and Boom 2007, p. 3). That is a fundamental idea in defining knowledge and using that definition for research purposes. It would be a mistake to take for granted a knowledge definition without understanding the supporting metaphor and its semantic limitations. Unfortunately, many researchers in knowledge management use knowledge definitions formulated by famous authors without checking for their metaphorical framework and their semantic limits. For instance, one of the most frequently cited working definition of knowledge has been formulated by Thomas Davenport and Laurence Prusak (2000, p. 5): "Knowledge is a fluid mix of framed experience, values, contextual information, and expert insight that provides a framework for evaluating and incorporating new experiences and information. It originates and is applied in the minds of knowers. In organizations, it often becomes embedded not only in documents or repositories but also in organizational routines, processes, practices, and norms". It is a descriptive definition that attempts to capture the main attributes of knowledge in an organizational context. Although we need such a working definition for knowledge, we should see the supporting metaphor and the limitations induced by it in using that concept of knowledge. In this particular case, the metaphor is that of *stocks and flows* which will be discussed in one of the following sections of this chapter.

1.2.2 Knowledge as Objects

The first class of metaphors developed by people who were in search for knowledge understanding and using it in practical organizational contexts is that of knowledge as *objects*, *stocks*, or *resources*. The explanation comes from the fact that objects are tangible with clear and easily identifiable attributes. In a research on the nature of intellectual capital and on the metaphors used by different authors, Andriessen (2006) shows that Davenport and Prusak used this kind of metaphors in the first chapter of their book *Working knowledge. How organizations manage what they know* in proportion of 59% of the total number of all metaphors used in that chapter, and Nonaka and Takeuchi used in chapter 5 of their book *The knowledge-creating company. How Japanese companies create the dynamics of innovation* (metaphors based on physical objects in proportion of 29% of the total number of metaphors used in that chapter). We provide these examples because both books have been very influential among all academics and practitioners involved in knowledge management and intellectual capital, and have contributed significantly to promoting knowledge metaphors based on physical objects and their attributes. The followings are just some examples of such metaphors, where we introduced italics to underline the main elements of these metaphors:

(1) "The idea of dealing with *knowledge as an object* has been already exploited in a variety of areas across knowledge management and information technology" (Borgo and Pozza 2012, p. 229).
(2) "A knowledge map can also serve as an inventory ... It therefore can be used as a tool to evaluate the corporate *knowledge stock*, revealing strengths to be exploited and gaps to be filled" (Davenport and Prusak 2000, p. 72).
(3) "The realization that *knowledge* is the new competitive *resource* has hit the West like a lightning" (Nonaka and Takeuchi 1995, p. 7).
(4) "Codification can be defined as a process of *storage*, *indexation* and *distribution* of formal *knowledge* independently of any context" (Janicot and Mignon 2012, p. 6).
(5) "Just as food and manufactured goods can be packaged and sold, there are ways *to package knowledge* for commercial benefit, using the intellectual property laws" (Sullivan 1998, p. 143).

The first example shows explicitly that knowledge should be understood in terms of an object, which means that the metaphor defines a framework with some structural and functional attributes coming from objects. That is confirmed by the second example where knowledge is considered to be like a stock, and the third example where knowledge is considered like a tangible resource in a company. The last two examples refer to the functional attributes of objects which have been transferred to the target domain. Thus, knowledge can be stored, indexed, distributed and packed like physical objects. Although these properties are very intuitive in describing knowledge, they induce the idea of considering knowledge like some individual entities which can be stored on a shelf, can be distributed like physical objects and can be subject to packaging operations like any commercial product. Some people may ask what is wrong with such a perspective or why we should be careful in treating knowledge this way. First, if we consider knowledge existing as individual entities like products in a supermarket which can be arranged on shelves and stored one upon the other, then we accept the idea of linearity and the summation operation. That leads to the idea of measuring the quantity of knowledge by counting the number of knowledge entities and performing the summation mathematical operation. Actually, this kind of attitude already exists and most of the metrics designed to evaluate knowledge and other intangible resources in organizations are linear metrics (see Chap. 8 for an additional discussion about this issue). Second, when distributing physical objects the initial quantity of them is progressively diminishing. In reality, when a person shares her/his knowledge with somebody else or disseminates it to a group of people, the initial quantity of knowledge does not diminish; it remains at the same level since knowledge is not composed of individual well-defined pieces which are removed from the initial inventory. Third, when physical objects are used frequently and for a long time, they are subjected to a degrading process. Knowledge can be used as much as we need it without any process of losing any of its properties. Just think of the Pythagorean theorem in mathematics or the Newtonian laws of physics.

These metaphors have been promoted mostly by researchers coming from information science and engineering who work with the Shannonian concept of *information*, which is devoid of any meaning (Bratianu 2015) and is a pure mathematical concept reflecting a certain distribution of probabilities. Due to its mathematical nature, this concept of information is objective, and its objectivity inspired some researchers to extend mathematical methods to the concept of knowledge and to find ways of its objectification. In this perspective, Bolisani, Borgo and Oltramari (2012, p. 203) remark that if "knowledge can be objectified, this means that it can be handled, reproduced, stored and transferred, largely independently from the individual that produces or possesses it". That objectified knowledge can be embedded into documents, software codes, databases, and different platforms for sharing it among the employees with a high probability of getting the same interpretation.

1.2.3 Knowledge Nuggets

The temptation of using simple and intuitive metaphors leads to the creation of the interesting expression of *knowledge nuggets*. From the well-known *chicken nuggets* you can order in McDonald's fast food restaurants, *knowledge nuggets* captured the imagination of IT experts who use it quite frequently in data processing, especially in data mining, knowledge discovery, and knowledge production processes (Carayannis and Campbell 2011; Delen and Al-Hawamdeh 2009; Williams and Huang 1997). According to the Oxford Advanced Learner's Dictionary (2004), the word *nugget* may have the following meanings: (a) a small lump of a valuable metal or mineral, especially gold, that is found in the earth; (b) a small round piece of some type of food: *chicken nuggets*; (c) a small thing such as an idea or a fact that people think of as valuable: *a useful nugget of information*. Thus, the concept of *knowledge nuggets* reflects the metaphorical thinking based on small and usually valuable objects. Also, it suggests an extension of the concept of *shannonian information* toward *semantic information*, although the first one is a mathematical concept without embedding any concrete meaning. The concept of *knowledge nuggets* leads intuitively to the idea of defining small pieces of information or knowledge which can be aggregated into larger structures, stored, retrieved, distributed and used. The exponential increase in data gathered and stored in huge databases generated a great conceptual effort to create new models and technologies for searching and retrieving useful information. In this context, Data Mining is "the process of identifying valid, novel, potentially useful, and ultimately understandable patterns in data stored in structured databases, where the data are organized in records structured by categorical, ordinal and continuous variables" (Delen and Al-Hawamden 2009, p. 142). By novel information, experts in data mining mean new correlations, trends, or patterns that can be discovered in the very large databases of the Big Data systems. The novel information is structured as knowledge nuggets which can be delivered to the interested users. The concept of knowledge nuggets is also used in e-learning and micro-learning programs, where it represents well-defined and meaningful structures of knowledge. Here, we have to make a clear distinction between the string of signs which corresponds to a knowledge nugget and the semantic content of that nugget. For example, we may consider as a knowledge nugget a trend found in a large database, expressed as a sentence. We may put together such sentences and sum them up into a paragraph. That is a linear operation applied to the strings of letters or to their digital correspondents which can be stored, retrieved, transferred or distributed. However, the meanings of nuggets cannot be aggregated on the same principle, since meaning is nonlinear and the result of such an aggregation may have no meaning at all. The conflicting situation is generated by the different significance of the concept of shannonian information used in computer science as a pure mathematical construct without any semantic content, and the concept of semantic information used in knowledge management.

In practical terms, knowledge nuggets can be the result of presenting some ideas, tips, rules, or practical suggestions very synthetically, by using both texts and images, like in a series of humorous videos for field sales agents which are posted on Youtube. Also, the Organization Migration4Development (M4D) uses knowledge nuggets as extractions of key concepts and ideas from projects, e-discussions, live chats and reports to inform the community with M4D issues at the local level. In a larger sense, knowledge nuggets may be conceived as a result of piecewise discretization process of a continuum of knowledge contained in a book, paper, program, conference or live chat and selection of most significant of them for the users. We can make a parallel with complex nonlinear phenomena in mathematics which cannot be solved as they are, and experts use different discretization methods to transform those continuum fields into discrete ones for which can be applied numerical methods to get useful solutions.

1.2.4 The Iceberg Metaphor

The iceberg metaphor has been used extensively by Ikujiro Nonaka and his colleagues since it is very simple and very intuitive for the conceptual dyad of explicit knowledge—tacit knowledge (Nonaka 1994; Nonaka et al. 2008; Nonaka and Takeuchi 1995; Nonaka and Von Krogh 2009). Indeed, as they recognize, the fundamental aspect of their epistemology is the distinction between explicit and tacit knowledge, distinction that can be easily understood by using the iceberg metaphor. "Thus, knowledge that can be expressed in words and numbers represents only the tip of the iceberg of the entire body of knowledge" (Nonaka and Takeuchi 1995, p. 60). Explicit knowledge is the rational knowledge that can be formulated by using any natural or symbolic language, and can be easily transferred in a social context. It is like the visible part of the iceberg. Tacit knowledge is personal knowledge and comes mostly from direct experience, which is processed by the cognitive unconscious. According to Nonaka and Takeuchi (1995, p. 8), "Tacit knowledge is personal and hard to formalize, making it difficult to communicate or to share with others. Subjective insights, intuitions, and hunches fall into this category of knowledge. Furthermore, tacit knowledge is deeply rooted in an individual's action and experience, as well as in the ideals, values, or emotions he or she embraces". Metaphorically, tacit knowledge can be represented by the hidden part of the iceberg. We know that it is there, under the water line, but we cannot see it and we have no idea how big that part of the iceberg is.

The iceberg metaphor captures our attention since it is simple and intuitive, but on the other hand it has serious limitations since the iceberg is a solid and there is no flow between its visible and hidden parts. Thus, there is no dynamics in the source domain which can be mapped onto the target domain to illustrate the conversion process of tacit knowledge into explicit knowledge. In addition, the split of the iceberg into two distinct parts can be only seen from an observer's perspective, since the iceberg is a homogeneous solid without any intrinsic differences between the upper and the lower parts. Thus, the distinction between tacit and explicit knowledge comes from a contextual attribute of the iceberg seen in the ocean's water and not from a material distinction between the upper and the lower parts, which questions the effectiveness of the structural mapping from the source domain onto the target domain of the metaphor.

1.2.5 Knowledge Flows

In order to eliminate the limitations introduced by the discrete nature of objects and their static forms, a new metaphor has been created by using the image of *fluid flows*. Thus, in the source domain, we have the semantic field associated to flow of fluids, while we have the semantic field of knowledge in the target domain. In the source domain, in a more advanced and complex metaphor, some authors consider both *stocks and flows* combining the attributes of the two semantic fields. Thus, *knowledge as stocks and flows* constitutes one of the most frequently used metaphors. Bolisani and Oltramari (2012, p. 280) explain the essence of this metaphor effectively: "We can denote knowledge stock as the amount or 'level' of knowledge possessed at a particular time in an organization, while knowledge flows identify knowledge that is transferred from one economic player to another. According to this interpretation, knowledge flows can affect the amount of knowledge stocked by the two players". We shall illustrate these metaphors with some examples taken from literature. We introduced italics for the metaphors used in the texts.

- "For this *flow of knowledge* to prevail, the organizational culture must be extraordinary" (Davenport and Prusak 2000, p. 109).
- "The way knowledge *flows* in organizations is often a hidden process" (O'Dell and Hubert 2011, p. 109).
- "Rapid and reliable *flows of knowledge* across people, organizations, times, and places are critical to enterprise performance. Unfortunately, the leader and manager have negligible current guidance for assessing and enhancing *knowledge flows* in practice. A dearth of contemporary research addresses the *dynamics of knowledge*, which are fundamental to understanding *knowledge flows*" (Nissen 2006, p. IX).
- "With the wider view I am taking, I claim that managing *knowledge flows* is something that can be applied and used in almost any type of organization" (Leistner 2010, p. 6).
- "So *flow of knowledge* from individuals depends on three broad factors: individual preferences, the social situation and organizational factors" (Oliver 2013, p. 19).

Fluid flows are well-known phenomena, easily to understand and explain. Unlike objects that have limited and well-defined geometries that are static and unchangeable in a uniform and constant field of forces, fluids have changeable geometries and have the property of flowing under the influence of a pressure field. They are dynamic. Fluids can be accumulated and stored in reservoirs, and distributed through channels or industrial piping systems. In nature, fluid flows in channels or rivers as a result of the gravity field, which means from a higher altitude to a lower one. In industry, cities or buildings, fluid flows through ducts and pipes from a higher pressure level created by a pump toward a lower pressure level. That motion of flow has been used many times in science to explain new phenomena like electrical current and heat flux. Even today, some people think that heat is flows from a hot physical object toward a cold one, and that electricity flows through a wire. Why not to consider that knowledge flows through an organizational structure from well-informed people toward less-informed ones?

Knowledge as *stocks and flows* is a complex metaphor composed of several simple ones which form analogies with fluids, their physical property of being a continuum and their functional attribute of flowing. Nissen (2006, p. XX) associated a fluid flowing through a piping system with knowledge flowing through an organizational structure: "To the extent that organizational knowledge does not exist in the form needed for application or at the place and time required to enable work performance, then *it must flow* from how it exist and where it is located to how and where it is needed. This is the concept *knowledge flows*". The model proposed by Nissen is an extension of the dynamic model developed by Nonaka and his colleagues (Nonaka 1994; Nonaka and Takeuchi 1995) since it is based on the SECI construct, but it contains *time* as a new dimension. While Nonaka's model is an inertial model, Nissen's one is really a dynamic model because it includes time. Nissen introduces two new dimensions: life cycle and flow time. "Life cycle refers to the kind of activity (e.g. creation, sharing, application) associated with knowledge flows. Flow time pertains to the length of time (e.g. minutes, days, years) required for knowledge to move from one person, organization, place, or time to another" (Nissen 2006, p. 35). It is useful to underline the fact that *knowledge flows* in Nissen's perspective refers not only to the motion of knowledge from one part of organization to another one, but also from one moment of time to another. Flow of time is important especially for intergenerational knowledge transfer and databases creation. However, Szulansky (1996, 2000) reveals that knowledge flows implies also *knowledge stickiness* manifested as a difficulty in the process of knowledge transfer. He says that knowledge can be *sticky*: "To a large extent, this is because internal transfer of knowledge, rather than fluid, is often 'sticky' or difficult to achieve" (Szulansky 2000, p. 10).

We have to observe that the metaphor *knowledge as objects* can be used only for explicit knowledge, while the metaphor *knowledge as stocks and flows* can be used for both explicit and tacit knowledge. Davenport and Prusak (2000, p. 5) used this metaphorical entailment in their famous definition: "Knowledge is a fluid mix of framed experience, values, contextual information and expert insight that provides a framework for evaluating and incorporating new experiences and information". However, the *knowledge flow* metaphor cannot explain the conversion of tacit knowledge into explicit knowledge which constitutes the essence of knowledge creation in Nonaka's model. Also, the metaphor is still based on the Newtonian physics which implies motion in space and linearity when dealing with knowledge. There is no transformation of phase or other type of changing the nature of the fluid to support the knowledge conversion processes postulated by Nonaka in his famous

SECI model. For overcoming these limitations we need to change the paradigm of Newtonian logic into the paradigm of entropic transformations as we shall explain in the next sections of this chapter.

1.3 The Energy Metaphor

1.3.1 Knowledge as Energy

In his seminal book *Corporate longitude: What you need to know to navigate the economy* (2002), Leif Edvinsson considers that we need to advance in understanding and explaining knowledge by developing new models and new metaphors. A possible new metaphor is *knowledge as energy* (Bratianu 2011, 2013, 2015; Bratianu and Andriessen 2008). In the source domain we consider *energy* with all its attributes, and in the target domain we consider *knowledge*. There are three main attributes or properties of energy we are interested to map onto the knowledge domain:

- Energy is a field.
- Energy manifests in different forms (i.e. mechanical, thermal, electrical etc.).
- One form of energy can transform into another form of energy. The transformation is irreversible.

The first attribute leads us to a new interpretation of knowledge which changes the main paradigm of its definition. Knowledge is not considered like a tangible object or a fluid flow anymore. It is considered like a field of forces which is intangible and forms a continuum both in space and time. For instance, we all are aware of the gravity field although we cannot see it and cannot touch it. But if we jump we feel immediately the attraction force of the earth. That means an intangible field of forces. Energy fields are usually distributed non-uniformly in space and have variations in time. These properties can be transferred to the knowledge field.

The second attribute is obvious for all of us. Energy can be found in nature in different existential forms like mechanical energy, thermal energy, electrical energy, nuclear energy etc. This attribute mapped onto the target domain leads to the idea that knowledge can manifest in different forms of different nature. The two forms discussed so far (i.e. tacit and explicit knowledge) are different not due to their nature but due to their way of being processed by our brain. Tacit knowledge is processed fundamentally by the unconscious zone of the brain, while the explicit knowledge is processed by the conscious zone of the brain where natural language plays an essential role. We can now consider three fundamental forms of knowledge: rational knowledge, emotional knowledge, and spiritual knowledge (Bratianu 2013, 2015). Rational knowledge is the result of the reasoning process and expresses concepts and ideas formulated in a natural or symbolic language. Rational knowledge is the explicit form of knowledge. Emotional knowledge is a wordless form of knowledge which is generated by our emotions and feelings. In

Nonaka's theory emotional knowledge is found in tacit knowledge mixed with spiritual knowledge which expresses our cultural values and ethical principles.

The third attribute comes from thermodynamics and reflects the capacity of energy to transform from one form into another in some given conditions. For instance, mechanical energy can transform through friction into heat. This attribute mapped from the source domain onto the target domain shows that one form of knowledge can transform into another form in given conditions. For instance, emotions of fear make us think of some protection or avoiding a dangerous situation. In such a context, emotional knowledge transforms into rational knowledge. These transformations are irreversible and they represent the content of the entropic knowledge dynamics, where entropy is a measure of irreversibility.

The energy metaphor allows us to propose a new paradigm for knowledge based on the multi-field theory of knowledge and the entropic knowledge dynamics. The multi-field theory says essentially that individual and organizational knowledge is represented by three fundamental fields of knowledge: rational, emotional, and spiritual. The entropic knowledge dynamics concerns the transformation of one form of knowledge into another in some given conditions. We shall present the main ideas of these new domains of research in the next sections.

1.3.2 The Field of Rational Knowledge

The multi-field theory of knowledge states that, at the individual level and organizational level, there are three co-existing fields of knowledge: rational, emotional, and spiritual. They are fundamental forms of knowledge manifestation which are generated and constituted in a different way. However, they are not independent fields but in a continuous interaction and transformation such that decision making incorporates contributions coming from all of them (Bratianu 2013, 2015). We may say that *knowledge* is a construct similar to the *white light* which can be decomposed in monochromatic lights when passing through a prism. That means that knowledge is an integrative concept containing rational, emotional, and spiritual knowledge. The new perspective is in concordance with the multiple intelligences model developed by Howard Gardner (1983, 2006). That model changed completely our idea that intelligence is a single entity which can be measured and expressed numerically by using the concept of *intelligence quotient (IQ)* created by Alfred Binet. Gardner defines intelligence as "a bio-psychological potential to process specific forms of information in certain kinds of ways. Human beings have evolved diverse information—processing capacities—I term these 'intelligences'—that allow them to solve problems or to fashion products" (Gardner 2006, p. 29).

The rational knowledge field contains rational knowledge which has been considered as the only form of knowledge for centuries by philosophers. We discussed about these epistemological aspects of knowledge in the beginning of the chapter. Rational knowledge is represented mainly by explicit knowledge since it is the result of the conscious cognitive brain. Descartes (1997, p. 147) expressed that conviction as follows: "Even bodies are not properly speaking known by the senses or by the faculty of imagination, but by the understanding only, and since they are not known from the fact that they are seen or touched, but only because they are understood. I see clearly that there is nothing which is easier for me to know than my mind". Rational knowledge is considered to be objective and this attitude made it suitable for developing scientific and technological knowledge. Also, education in the western countries has been conceived in objective terms and by stressing the importance of science and technology which means the primacy of rational knowledge. Rational knowledge is framed into explicit knowledge by using a natural or symbolic language: "Language serves not only to express thoughts, but to make possible thoughts which could not exist without it" (Russell 1992, p. 58). Organizational rational knowledge is obtained by integrating all individual rational knowledge fields and all documents and databases which contain data, information, and knowledge. Classical decision making theory is based on rational knowledge and expressed mostly in the symbolic language of mathematics. Knowledge management has been developed, in its first phase, on rational knowledge as an extension of the information management which is centered on the concept of shannonian information and information technology. That is why managers developed their generic strategies based on rational knowledge and information technology.

1.3.3 The Field of Emotional Knowledge

The emotional knowledge field contains knowledge generated by emotions and feelings. Emotional knowledge is a wordless form of knowledge which is processed by the unconscious part of our brain. Emotional knowledge is generated in the direct contact of our body with the external world, and is integrated into what we call experience. Also, emotional knowledge can be obtained by processing information coming from our internal body. Emotional knowledge emerged as a component of tacit knowledge, especially after the work of Michael Polanyi (1983). In his seminal book about the tacit dimension of knowledge, Polanyi considers our direct experience with the environment as a source of knowing. It is a bodily experience which generates emotional information through perception, information which becomes then emotional knowledge. "I said that by elucidating the way our bodily processes participate in our perceptions we will throw light on the bodily roots of all thought, including man's highest creative powers" (Polanyi 1983, p. 15).

Human resources management demonstrated that emotional knowledge plays a crucial role in motivating people for working very hard and achieving performance. Motivation becomes critical during change processes when there is a need for greater efforts without immediate rewards. Understanding and using emotional knowledge in influencing people makes the difference between managers and leaders, since managers prefer numbers and rational decisions while leaders influence people acting on their emotional and spiritual knowledge fields. John Kotter, who studied organizational change and leadership involved in performing them, demonstrated that in any change process emotional knowledge is much more important than rational knowledge. Kotter showed that analytics could be interesting, but not always convincing. For example, rational knowledge is needed for understanding the logic of change but could be not enough for changing employees' behavior. Much more convincing could be for them to feel the need of change as a result of emotional knowledge transferred to them by leaders. "The single biggest challenge in the process is changing people's behavior. The key to this behavioral shift, so clear in successful transformations, is less about analysis and thinking and more about seeing and feeling" (Kotter and Cohen 2002, p. 179). In change management, the old paradigm of *analyzing-thinking-changing* should be replaced with new one of *seeing-feeling-changing*. Thus, the action of *seeing* creates the perceptions able to generate through *feeling* the necessary emotional knowledge needed to contribute together with rational knowledge to *changing* people's behavior. That means that emotional knowledge contributes significantly to the decision making both at individual and organizational levels. As Dan Hill (2008, p. 2) remarks, "Breakthroughs in science have revealed that people are primarily emotional decision makers". Based on this idea and many psychological investigations of decision making processes, Malcolm Gladwell introduces, in his famous book *Blink*, the concept of "thin-slicing" decision making: "Thin-slicing refers to the ability of our unconscious to find patterns in situations and behavior based on very narrow slices of experiences" (Gladwell 2005, p. 24). Many people say this is *intuition*, since intuition is a result of our condensed and filtered experience powered by emotional intelligence. These two fields of knowledge are related to the multiple intelligences structure of our thinking. In a synthetic way, Daniel Kahneman (2011) explains the fact that people developed during the history of humanity two modes of thinking that are interacting dynamically: (1) the emotional system that operates automatically and quickly, with almost no effort or sense of voluntary control, and (2) the rational system that operates slowly due to the many computations and choices it does. While the classical management theory, in its effort of proving that is a science, ignores the function of the first system because of its subjectivity, knowledge management considers both of them. Looking at the literature in this domain, we may say that authors coming from western countries are still emphasizing the role of rational system while authors coming from Japan emphasize the emotional system and tacit knowledge.

1.3.4 The Field of Spiritual Knowledge

Spiritual knowledge has been included by Nonaka and Takeuchi in tacit knowledge, mixed up with emotional knowledge (Nonaka and Takeuchi 1995). We consider spiritual knowledge essential for our existence, fact for which we introduce it as a fundamental field in the new multi-field theory of knowledge. Spiritual knowledge integrates values and beliefs about life and about our existence and represents the backbone of spiritual capital of any organization (Zohar and Marshall 2000, 2004). "Our spiritual capital is our shared meaning, our shared purpose, our shared vision of what most deeply matters in life—and how these are implemented in our lives and in our behavioral strategies. It is the capital that is increased by drawing on the resources of the human spirit" (Zohar and Marshall 2004, p. 27). If rational knowledge reflects the objectivity of the physical environment we are living in, and emotional knowledge reflects the subjectivity of our body interaction with the external world, spiritual knowledge reflects our understanding about the meaning of our existence. As Maxwell (2007, p. 274) states, "We have to learn to see aspects of the world around us: stones, people, trees, sky. Equally, we have to learn to see meaning and value in the world around us, in our environment, in events, in human actions and lives".

Individuals working together in a company share their values and beliefs about life, work and future generating in time an organizational culture and an working spirituality. It is a way of thinking and feeling inextricably connected with caring, hope, kindness, love and optimism. Spiritual knowledge is essential in decision making since rational arguments are strongly influenced by the value settings. We are all aware of the fact that positive values correlate directly with the business success, while negative values lead managers toward business failures. Thus, spiritual knowledge which reflects positive values and positive spiritual intelligence is essential in conceiving successful strategies and in achieving competitive advantage. Spiritual knowledge is intrinsically related to the concept of Corporate Social Responsibility, a concept requesting responsible governance and a vision driven by social values and not profit maximization (Basu and Palazzo 2008; Branson 2011; Pinto et al. 2008; Wang et al. 2011).

1.4 Conclusion

Knowledge is a universal concept which attracted the attention of philosophers from ancient times. There were countless efforts to define it following the rules of scientific inquiry, but the resulting definitions were not able to integrate all the semantic attributes of knowledge. Searching for an objective perspective and a rational approach many philosophers eliminated all subjective aspects related to perception and bodily involvement claiming that knowledge is *a justified true belief*. However, the precision and logical coherence, used in the theoretical approach to knowledge, generated uncertainty in the practical modalities of justifying the truth. If we agree with Nonaka and Takeuchi (1995, p. 87) that "justification criteria need not be strictly objective and factual", then the philosophical meaning of truth is almost lost. Truth and its justification cannot have the same degree of objectivity anymore. We may think of Heisenberg's uncertainty principle used in quantum mechanics that states, in the case of nuclear particles, that position and velocity cannot be measured exactly at the same time. Knowledge is created by human brain and then it is amplified and integrated into organizational knowledge by social interaction. That means that knowledge comprises both objective and subjective attributes. Objective attributes can be conceived as being independent of the social context, but the subjective attributes are context dependent and cannot be transferred easily to some other similar contexts. Knowledge sharing can be a good example for such kind of situations.

Cognitive scientists demonstrated that our mind works metaphorically. That means that we use metaphors to understand and explain a less known concept or experience in terms of other well-known one. Conceptual metaphors have a simple structure composed of a source domain where we place the well-known concept and a target domain where we place the new or less known concept. By using structural mapping, some of the main attributes of the concept framed within the source domain are transferred to the concept put in the target domain, enlarging this way its semantic field. Since knowledge is an abstract concept without any reference to some tangible objects, authors use explicit or implicit metaphors in dealing with it and with knowledge management. The first class of metaphors developed for knowledge explanation is based on those that contain physical objects with tangible attributes in the source domain. It is the favorite class of metaphors used by authors dealing with knowledge as strategic resources. Thus, knowledge can be accumulated, stored, distributed, packed and delivered like tangible objects. From that class derived the iceberg metaphor which has been used extensively to explain the pair of explicit and tacit knowledge. Knowledge nuggets are an extension of the same category of metaphors, but resulted from a discretization of a continuum of knowledge (i.e. a text containing a narration or a story). The most advanced class of metaphors are those based on stocks, flows, or stocks and flows. Thus, knowledge is conceived like a fluid flowing through organizations, from where it is created to where it is needed.

All of these metaphors presented above induce a series of limitations in understanding and using the full potential of knowledge. These limitations derive from the Newtonian logic, the linearity property and the illusion of measuring knowledge by using the methods developed for tangible objects and their attributes. In an effort to overcome these limitations, a new metaphor based on energy is proposed in Bratianu and Andriessen (2008). According to this new perspective, knowledge is conceived like a field without any tangible attributes. Moreover, following the analogy with the co-existence of multiple forms of energy (i.e. mechanical, thermal, electrical, nuclear etc.), the existence of three fundamental fields of knowledge is postulated: rational, emotional, and spiritual. Rational knowledge is basically explicit knowledge since it is framed by our reasoning mind and natural language. It is a construct following the Cartesian spirit. Emotional knowledge is a wordless expression of our body response to the external environment and it is a direct result of emotions and feelings. Emotional knowledge is subjective and context dependent. Spiritual knowledge contains values and ethical principles and is essential in decision making. Both emotional and spiritual knowledge have been embedded in tacit knowledge and mixed up in the fuzzy description of experience. The energy metaphor constructs a new paradigm which allows us to have a better understanding of knowledge and to offer managers and leaders new opportunities to influence people in times of change and uncertainty.

References

Andriessen, D. (2006). On the metaphorical nature of the intellectual capital: A textual analysis. *Journal of Intellectual Capital, 7*(1), 93–110.

Andriessen, D., & Boom, M. (2007, May). *Asian and Western intellectual capital in encounter*. Paper presented at IC-Congress 2007, Inholland University of Applied Sciences, Haarlem, The Netherlands.

Ayer, A. J. (2009). The right to be true. In R. Neta & D. Pritchard (Eds.), *Arguing about knowledge* (pp. 11–13). London: Routledge.

Basu, K., & Palazzo, G. (2008). Corporate social responsibility: A process model of sensemaking. *Academy of Management Review, 33*(1), 122–136.

Bolisani, E., & Oltramari, A. (2012). Knowledge as a measurable object in business contexts: A stock-and-flow approach. *Knowledge Management Research and Practice, 10*(3), 275–286.

Bolisani, E., Borgo, S., & Oltramari, A. (2012). Using knowledge as an object: Challenges and implications. *Knowledge Management Research & Practice, 10*(3), 202–205.

Borgo, S., & Pozza, G. (2012). Knowledge objects: A formal construct for material, information and role dependences. *Knowledge Management Research and Practice, 10*(3), 227–236.

Branson, R. (2011). *Screw business as usual*. London: Virgin Books.

Bratianu, C. (2011). Changing paradigm for knowledge metaphors from dynamics to thermodynamics. *Systems Research and Behavioral Sciences, 28*(2), 160–169.

Bratianu, C. (2013). The triple helix of organizational knowledge. *Management Dynamics in the Knowledge Economy, 1*(2), 207–220.

Bratianu, C. (2015). *Organizational knowledge dynamics: Managing knowledge creation, acquisition, sharing, and transformation*. Hershey: IGI Global.

Bratianu, C., & Andriessen, D. (2008). Knowledge as energy: A metaphorical analysis. In D. Harorimana & D. Watkins (Eds.), *Proceedings of the 9th European conference on knowledge management* (pp. 75–82). Reading: Academic Publishing.

Carayannis, E. G., & Campbell, D. F. J. (2011). Open innovation diplomacy and the 21st century fractal research, education and innovation (FREIE) ecosystem: Building on the quadruple and quintuple helix innovation concepts and the "mode 3" knowledge production. *Journal of Knowledge Economy, 2*, 327–372.

Davenport, T. H., & Prusak, L. (2000). *Working knowledge. How organizations manage what they know*. Boston: Harvard Business School Press.

Delen, D., & Al-Hawamdeh, S. (2009). A holistic framework for knowledge discovery and management. *Communications of the ACM, 52*(6), 141–145.

Descartes, R. (1997). *Key philosophical writings*. Hertfordshire: Wordsworth Editions.

Dombrowski, E., Rotenberg, L., & Bick, M. (2013). *Theory of knowledge*. Oxford: Oxford University Press.

Edvinsson, L. (2002). *Corporate longitude: What you need to know to navigate the knowledge economy*. London: Prentice Hall.

Gardner, H. (1983). *Frames of mind: The theory of multiple intelligences*. New York: Basic Books.

Gardner, H. (2006). *Changing minds: The art and science of changing our own and other people's minds*. Boston: Harvard Business School Press.

Gentner, D., Bowdle, B. F., Phillip, W., & Borant, C. (2001). *The analogical mind. Perspectives from cognitive science* (pp. 199–254). Cambridge: The MIT Press.

Gettier, E. (2009). Is justified true belief knowledge? In R. Neta & D. Pritchard (Eds.), *Arguing about knowledge* (pp. 14–15). London: Routledge.

Gladwell, M. (2005). *Blink. The power of thinking without thinking*. New York: Back Bay Books.

Hill, D. (2008). *Emotionomics. Leveraging emotions for business* (Rev. ed.). London: Kogan Page.

Janicot, C., & Mignon, S. (2012). Knowledge codification in audit and consulting firms: A conceptual and empirical approach. *Knowledge Management Research & Practice, 10*(1), 4–15.

Kahneman, D. (2011). *Thinking fast and slow*. New York: Farrar, Straus and Giroux.

Kaufman, S. F. (1994). *The martial artist's book of five rings. The definitive interpretation of Miyamoto Musashi's classic book of strategy*. Boston: Tuttle Publishing.

Klein, G. (2003). *The power of intuition: How to use your feelings to make better decisions at work*. London: Currency/Doubleday.
Kotter, J. P., & Cohen, D. (2002). *The heart of change: Real-life stories of how people change their organizations*. Boston: Harvard Business School Press.
Lakoff, G. (1990). The invariance hypothesis: Is abstract reason based on image-schema? *Cognitive Linguistics, 1*(1), 39–74.
Lakoff, G., & Johnson, M. (1999). *Philosophy in the flesh. The embodied mind and its challenge to Western thought*. New York: Basic Books.
Lehrer, K. (2009). Knowledge, truth and evidence. In R. Neta & D. Pritchard (Eds.), *Arguing about knowledge* (pp. 16–21). London: Routledge.
Leistner, F. (2010). *Mastering organizational knowledge flow: How to make knowledge sharing work*. Hoboken: Wiley.
Maxwell, N. (2007). *From knowledge to wisdom: A revolution for science and humanities* (2nd ed.). London: Prentice Hall.
Neta, R., & Pritchard, D. (2009). *Arguing about knowledge*. London: Routledge.
Nissen, M. E. (2006). *Harnessing knowledge dynamics: Principled organizational knowing & learning*. London: IRM Press.
Nonaka, I. (1994). A dynamic theory of organizational knowledge creation. *Organization Science, 5*(1), 14–37.
Nonaka, I., & Takeuchi, H. (1995). *The knowledge-creating company: How Japanese companies create the dynamics of innovation*. New York: Oxford University Press.
Nonaka, I., & Von Krogh, G. (2009). Tacit knowledge and knowledge conversion: Controversy and advancement in organizational knowledge creation theory. *Organization Science, 20*(3), 635–652.
Nonaka, I., Toyama, R., & Hirata, T. (2008). *Managing flow: A process theory of the knowledge-based firm*. Houndmills: Palgrave Macmillan.
O'Dell, C., & Hubert, C. (2011). *The new edge in knowledge: How knowledge management is changing the way we do business*. New York: Wiley.
Oliver, G. (2013). A tenth anniversary assessment of Davenport and Prusak (1998/2000) working knowledge: Practitioner approaches to knowledge in organizations. *Knowledge Management Research & Practice, 11*(1), 10–22.
Pinker, S. (2008). *The stuff of thought. Language as a window into human nature*. New York: Penguin Books.
Pinto, J., Leana, C. R., & Pil, F. K. (2008). Corrupt organizations or organizations of corrupt individuals? Two type of organizational-level corruption. *Academy of Management Review, 33* (3), 685–710.
Polanyi, M. (1983). *The tacit dimension*. Gloucester: Peter Smith.
Russell, B. (1972). *A history of Western philosophy*. New York: Simon and Schuster.
Russell, B. (1992). *Human knowledge: Its scope and limits*. London: Routledge.
Sullivan, P. H. (1998). *Profiting from intellectual capital: Extraction value from innovation*. New York: Wiley.
Szulansky, G. (1996). Exploring internal stickiness: Implements to the transfer of best practice within the firm. *Strategic Management Journal, 17*, 27–43.
Szulansky, G. (2000). The process of knowledge transfer: A diachronic analysis of stickiness. *Organizational Behavior and Human Decision Processes, 82*(1), 9–27.
Wang, L., Malhotra, D., & Murnigham, J. K. (2011). Economics education and greed. *Academy of Management Learning & Education, 10*(4), 643–660.
Williams, G. J., & Huang, Z. (1997, December). *Mining the knowledge mine: The hot spot methodology for mining large real world databases*. Paper presented at the 10th Australian Joint Conference on Artificial Intelligence, Perth, Western Australia.
Zohar, D., & Marshall, I. (2000). *SQ: Spiritual intelligence. The ultimate intelligence*. London: Bloomsbury.
Zohar, D., & Marshall, I. (2004). *Spiritual capital. Wealth we can live by*. San Francisco: Berrett-Koehler.

2 The Emergence of Knowledge Management

The purpose of this chapter is to show that knowledge management emerged as a necessity in the post-industrial society and the new knowledge economy. Instead of starting from defining knowledge management and describing its functions to create a prescriptive framework, the chapter begins with the broad picture of the changes in the structure of economy and in its critical assets. These changes produced a new type of economy where scarcity of tangible resources has been replaced by the affluence of intangible resources, and the economic theories of resource optimization and profit maximization have been aligned to knowledge creation and business sustainability. The engine of knowledge economy is the knowledge-based organization, where the pressure of efficiency and productivity should be relaxed. Instead, there is a need to develop new metrics able to measure the quality of knowledge and to evaluate the contribution of organizational learning to the firm's performance. Finally, the chapter presents the new attributes of knowledge workers and knowledge processes. Knowledge creation, acquisition, storing and retrieving, sharing and distribution, transformation and use become the components of knowledge management. Since knowledge and its functions constitute strategic resources, knowledge management bridges the gap between operational management and strategic management.

2.1 The Knowledge Economy

2.1.1 The Fourth Discontinuity

In his seminal book *The age of discontinuity: Guidelines to our changing society* published in 1969, Peter Drucker reveals some major changes in four areas of the world economy with great impacts on the whole society. He calls these major changes *discontinuities* which "while still below the visible horizon, are already changing structure and meaning of economy, polity, and society" (Drucker 2008, p. xxvii). He considers that these discontinuities, aggregated on their main

E. Bolisani, C. Bratianu, *Emergent Knowledge Strategies*, Knowledge Management and Organizational Learning 4, DOI 10.1007/978-3-319-60657-6_2

dimensions, result in a "recent future" since they show some accomplished facts and some challenges to come at the same time. They may play the role of an "early-warning system" for decision makers in all areas of the modern society. The major discontinuities identified by Peter Drucker are in the following areas:

- The emergence of new technologies, able to generate new industries and businesses.
- The transition from the "international economy" to the "world economy" which finally leads to the creation of the one global market.
- The political matrix of social and economic life, which is changing fast and leads to a new sociopolitical reality.
- The emergence of a knowledge economy. "Knowledge, during the last few decades, has become the central capital, the cost center, and the crucial resource of the economy. This changes labor forces and work, teaching and learning, and the meaning of knowledge and its politics" (Drucker 2008, p. xxix).

Drucker recognizes our difficulties in forecasting the future since human mind has always been focused on continuity, which means to extrapolate today's reality into the immediate future. By evaluating the past developments in industry and economy, experts can measure results and anticipate some trends based on them. But they cannot anticipate discontinuities and predict crises into a future that refuses to align to the living present.

Inspired by the Princeton's economist Fritz Machlup who coined the term "knowledge industries" in his book *Production and distribution of knowledge in the United States* (1962), Drucker coined the concept of "knowledge economy" and used it as the title of Chapter 12 in his book. Using statistical data, he shows how knowledge, knowledge work, and knowledge workers became significant phenomena in the new economies of developed countries like United States. As Drucker (2008, p. 264) remarks, statistics "do not reveal the important thing. What matters is that knowledge has become the central 'factor of production' in an advanced, developed economy". "Knowledge" is somehow different than "knowledge" used by intellectuals since it is linked to action. While the intellectuals' knowledge is theoretical and used in books and scientific journals, organizational knowledge is actionable and used by knowledge workers in their practice. From this point of view, Drucker considers that the most important step toward a knowledge economy has been done by Frederick W. Taylor in his pioneering scientific management. By introducing engineering methods in organizing and measuring the efficiency of industrial work, he was able to increase almost exponentially work productivity. That was possible not by working harder, but by working smarter. Thus "The key to productivity was knowledge, not sweat" (Drucker 2008, p. 271).

In his visionary book *The coming of post-industrial society* (1973) Daniel Bell dedicated a full chapter to the knowledge dimensions and the role of new technologies in structuring the new class of post-industrial society. Being inspired by the same book written by Fritz Machlup, Bell defines knowledge as "a set of organized statements of facts or ideas, presenting a reasoned judgment or an experimental result, which is transmitted to others through some communication medium in some systematic form" (Bell 1999, p. 175). The rationalistic perspective adopted by Bell and his social setting for knowledge transfer and use in the post-industrial society is obvious. In demonstrating the increasing role played by knowledge in the new society, Bell focused his attention on scientific knowledge published in books and journals, and less on knowledge embedded in products and services. Although there is a different perspective in designing the future of the "information age" than that used by Drucker, the message reveals the same significance: the coming of a new economy in which the dominant capital is not land, labor or money anymore. It is knowledge.

2.1.2 Key Features of the Knowledge Economy

Drucker's vision of the coming knowledge economy is now a reality. Knowledge became the driving force of knowledge economy, and knowledge economy became the engine of global economy growth. Powell and Snellman (2004, p. 199) define knowledge economy as "production and services based on knowledge intensive activities that contribute to an accelerated pace of technical and scientific advance, as well as rapid obsolescence". Thus, it is an economy in which knowledge is created, acquired, transformed, transferred, disseminated and used effectively to enhance economic development. A key feature of the knowledge economy is a greater reliance on intangible resources and on intellectual competences of using them, than on physical resources. A car today represents less a metal manufactured product and more a smart machine that incorporates data, information, knowledge and a lot of intelligence. Same remark can be made of many other products that changed completely our way of thinking in producing and using them.

The knowledge economy is not about scarcity of resources but about their affluence. All tangible resources that contribute to achieving a competitive advantage for a firm are scarce and using them in the production process should be done efficiently. Intangible resources like data, information and knowledge are not scarce anymore. On the contrary, in many domains there is an overflow of them. From scarcity of natural resources, we reach a stage of abundance of intangible resources which have the property of being non-rival assets. They can be deployed at the same time in multiple uses, and they don't decrease by dissemination. The real problem now is not the quantity of data, information and knowledge, but the capacity of analyzing and interpreting them. Even with the new technology of Big Data and the intelligent programs of data mining, the human capacity of processing the meanings and making decisions in conditions of uncertainty is still very limited (Davenport 2005; North and Gueldenberg 2011; Spender 2014). Thus, knowledge economy is about abundance of intangible resources and limited human capacity of processing them.

The ownership of knowledge assets is also debatable since knowledge is created in the minds of people and only a part of it can be aggregated and transformed into organizational knowledge which can be owned and controlled by the firm. That changes completely the way in which knowledge assets can be managed and used in the production process of goods and services (Dalkir 2005; Jashapara 2011; Nonaka and Takeuchi 1995). Tacit knowledge which integrates employees' experience can be lost when they retire or just leave the firm for some other jobs. Also, controlling knowledge assets cannot be done in the same way as for tangible resources since knowledge creation, sharing and dissemination depend strongly on individual capacity of performing these activities and on willingness of employees of doing them. Thus, in the knowledge intensive organizations, the rewarding systems switch from external to internal stimulating factors based on phenomena which can be described by the new multi-field theory of knowledge and the new knowledge dynamics we have presented in Chap. 1. Motivating people is a complex process based especially on the transformation of emotional knowledge into rational knowledge and spiritual knowledge, which is generally ignored in traditional industrial management that grounds on rationalistic decision making and economic efficiency (Zohar and Marshall 2004).

In the industrial economy, growth can be obtained by linear accumulation of tangible assets and their combination in the production process. In the knowledge economy, linear accumulation is not possible since intangible assets are nonlinear and they should be integrated, which is a totally different approach. Nonlinearity (Bratianu 2009) implies new metrics of evaluation and reporting of intellectual capital. In the knowledge economy we deal with different types of intangibles which request different metrics and scales leading frequently to the incommensurability problem (Spender 2014; see also Chap. 8 for details). Even the well-known scales of space and time change as a result of information technology. For global markets, the geography of production plants is losing its importance, and time is shrinking for the knowledge cycles and innovation processes. Speed of change and innovation strategies are the new characteristics of business models. "Though optimization, as a process, is equally important in the knowledge economy, it alone cannot create or maximize value. The only way to create value in the knowledge economy is by adopting innovation as the core business process. An organization's ability to create value depends on its innovation process, its intellectual resources, and the creativity of its workforce—its intellectual assets" (Al-Ali 2003, p. 8).

The knowledge economy became possible as a result of structuring a new social class of people able to contribute with their creativity to the new wealth of society. Richard Florida (2002, p. 68) demonstrates that the knowledge economy is intrinsically related to the development of a *creative class* which "consists of people who add economic value through their creativity. It thus includes a great many knowledge workers, symbolic analysts and professional and technical workers, but emphasizes their true role in the economy". That may constitute also an argument against those authors who equate knowledge economy with the powerful new information technology, since the centrality of the creative class represents the crucial role played by human capital in shaping that economy.

Information technology does not create knowledge but constitutes the basic support of processing data and information at amazing speed, and of communicating them in complex networks. Also, the new information platforms enable knowledge seeking and knowledge sharing within the boundaries of firms or across them. Information technology enables creating open innovation systems and virtual communities of practice (O'Dell and Hubert 2011). However, the efficiency of using information technology depends on the managerial capacity of each firm and on some necessary structural changes to accommodate the requirements of that technology. Also, to use efficiently new information technologies and their applications, people should be trained and helped in understanding their logic and smart operation. As Powell and Snellman (2004, p. 208) emphasize, "the long-expected gains in productivity from investments in information technology are not fully realized until complementary institutional arrangements are developed".

In the knowledge economy, new interactions between social institutions can be generated for stimulating knowledge production, transfer and application. For instance, Leydesdorff (2006) analyzes the importance of developing a triple helix of university-industry-government relations. The role of the government is to create a stimulating legislation framework for developing such new structures and clusters of science and technology, where universities contribute to knowledge production and transfer toward industry, and knowledge is used for producing new goods and services. "In other words, university-industry-government relations develop in terms of institutional arrangements that recombine three functions of the socio-economic system: (1) wealth generation and retention, (2) novelty production, and (3) control at the interfaces of these sub-dynamics" (Leydesdorff 2006, p. 48).

The knowledge economy created a new market able to deal with knowledge and knowledge embedded in goods and services. Knowledge is not only a resource but also a product itself that can be the object of a market transaction. People are knowledge seekers because they have to solve problems in conditions of uncertainty and incompleteness information. Thus, they are potential buyers of information and knowledge to solve new problems or to create new goods and services. As Davenport and Prusak (2000, p. 25) remark, "Like markets for goods and services, the knowledge market has buyers and sellers who negotiate to reach a mutually satisfactory price for the goods exchanged. It has brokers who bring buyers and sellers together and even entrepreneurs who use their market knowledge to create internal power bases".

2.2 The Knowledge Organization

2.2.1 The Economic Theory of the Firm

The emergence of *Knowledge Management* cannot be understood and explained without a deep understanding of the nature and functioning of the firm. As Spender (2015a, p. 4) emphasizes, "it is crucial to see KM cannot and does not embrace the entirety of human knowing. It always hinges on a 'theory of the firm', a boundary concept that separates organizational knowing from broader epistemological matters". Many economists took it for granted that firms came into existence as a possible answer to the complexity of the new enterprises created by the industrial revolution and made no effort to explain their nature and purpose. In a seminal paper on this topic, Coase (1937) demonstrated that it is necessary to explain the ideal concept of *the firm* and to show its difference from a real world firm. Also, he considered necessary to build a theory able to reveal the determinants of firm growth and its behavior with respect to the market. He analyzed the main approach developed by economists, based on the price mechanism, and found it inconsistent with the economical and managerial phenomena that occur within the firm. He shows that if the price mechanism is considered the fundamental principle based on which all decisions should be made within an economic system, then we cannot have a rational explanation for resource allocations within a firm and especially for the motivation of an entrepreneur in coordinating activities in a real world firm. As a consequence, Coase wanted to bridge the gap between the economic theory of resource allocation based on the price mechanism at the macroeconomic level, and the theory of resource allocation within a firm where decisions are made by the entrepreneur. In his view, "A firm, therefore, consists of the system of relationships which comes into existence when the direction of resources is dependent on an entrepreneur" (Coase 1937, p. 393). Based on that assumption, Coase goes on analyzing the economic factors that determine the size of the firm, especially the costs associated with the increasing complexity of managerial decision making under uncertainty, an issue introduced by Knight (1921). Uncertainty means that knowledge about the future is incomplete and that events do not have a deterministic behavior. In these conditions, managerial decisions always involve some risks and a bounded rationality approach, as demonstrated by Simon (1976).

Edith Penrose, in her landmark book *The theory of the growth of the firm*, whose first edition has been published in 1959, explains that "The 'theory of the firm'—as it is called in the literature—was constructed for the purpose of assisting in the theoretical investigation of one of the central problems of economic analysis—the way in which prices and the allocation of resources among different uses are determined" (Penrose 2013, p. 11). She found that previous attempts to formulate a theory of the firm were insufficient for the complexity of processes within the firm and powerless in revealing the forces able to contribute to the internal growth of the firm. In fact, a theory of the firm is a conceptual model to represent, at a certain abstraction level, the main processes within a firm. Consequently, the conclusions of any analysis based on the model should be validated then by the concrete facts within real world organizations. Penrose identifies, as relevant aspects of a firm, its administration and all resources used as inputs in the production process of goods and services. "Thus, a firm is more than an administrative unit; it is also a collection of productive resources the disposal of which between different uses and over time is determined by administrative decisions" (Penrose 2013, p. 24). An interesting remark she makes is on the potential services that resources may have. That distinction between resources and their potential services leads to the different ways firms are deploying their resources, and as a consequence to the creation of the competitive advantage on the market.

Since this book is about strategic thinking and knowledge strategies, it is worth to see how Penrose made a clear distinction between business opportunities and their perception by decision makers: "Although the 'objective' productive opportunity of a firm is limited by what the firm is able to accomplish, the 'subjective' productive opportunity is a question of what it thinks it can accomplish. 'Expectations' and not 'objective facts' are the immediate determinants of a firm's behavior, although there may be a relationship between expectations and 'facts'—indeed there must be if action is to be successful" (Penrose 2013, p. 41).

In her analysis about the role of resources in the growth of the firm, Penrose shows the importance of knowledge in its explicit and tacit forms, about 30 years before the Nonaka's analysis. She makes the difference between explicit knowledge which is "objective" and can be communicated, and the other form which is "subjective" and integrates the personal experience. The objective knowledge is about things and reflects the "know that" or "know about" while the other form of knowledge reflects "know how" of doing something, as demonstrated later by Ryle (1949). Experience is a powerful source of knowledge since it "produces increased knowledge about things and contributes to 'objective' knowledge in so far as its results can be transmitted to others. But experience itself can never be transmitted; it produces a change—frequently a subtle change—in individuals and cannot be separated from them" (Penrose 2013, p. 53). This idea is very important for understanding her theory of the firm and will be developed later by other authors in their theories about knowledge-based firms.

Penrose's analysis goes beyond the boundary of the firm and considers that knowledge coming from the external business environment can also be useful: "Clearly external changes may also become part of a firm's 'stock of knowledge' and consequently they may change the significance of resources to the firm. Knowledge of markets, of technology being developed by other firms, and of the tastes and attitudes of consumers, are of particular importance" (Penrose 2013, p. 79).

Replacing the economic perspective with a sociological one, Simon (1991, p. 3) defines an organization in terms of roles played by its members: "organizations are best viewed as systems of interrelated roles, and that is the way I have been viewing them here". He completes that definition by emphasizing that roles are not prescribed behaviors but a system of prescribed decision premises. "Roles tell organization members how to reason about the problems and decisions that face them: where to look for appropriate and legitimate informational premises and goal (evaluative) premises, and what techniques to use in processing these premises" (Simon 1991, p. 3).

The theory of the firm is essentially a modeling approach to extract the main variables and the correlations they create from the real world firms and to construct an abstract view of that reality. Economists were interested mostly in the ownership and control of tangible resources for increasing shareholders' profit through their efficient allocation and use. However, as Spender critically analyzes, none of those theories are "entirely clear what the firms envisioned are supposed to be doing, where their boundaries lie, what goals they pursue, or how they might be analyzed or evaluated—against maximizing profit, serving the community, optimizing cost and benefit allocation between stockholders, seeking sustained competitive advantage, or simply surviving?" (Spender 2015b, p. 57). These were some of Coase's (1937) "killer questions" and in Spender's view they are still without adequate answers. Spender rejects the reductionist approach that a firm is a bundle of economic resources, or political relationships. For him, a firm represents a complex socio-economic patterns emerged as "context-contingent constructed responses to the uncertainties that cause social anxiety" (Spender 2015b, p. 58).

Firms exist as a consequence of some entrepreneurs who aggregated some strategic inputs for value creation. That means to go beyond data computation and rationalistic decision making and reach the complexity of human nature. Managers are not limited in their problem solving to the rational methods and objective facts, and are also capable of using their non-rational experience, intuition and imagination in decision making. Judgment would enlarge their operational field and imagination will help them in dealing with absence of knowledge. Spender's theory of managed firm introduces Simon's concept of bounded rationality and Knight's uncertainty as main constraints in managerial decision making and reveals the fundamental role of *knowledge absence* in that process. "My thesis is that value is created as a manager—an entrepreneur, business leader or strategist—uses their imagination to reach into a confronted knowledge-absence, thereby projecting their judgment into the lived world and thereby agentically transforming it" (Spender 2015b, p. 59). Thus, the theory of the firm should incorporate this driving force of searching for practical ways of reducing knowledge-absence caused by knowledge incompleteness, indeterminacy and incommensurability (Spender 2014). Emergence of knowledge management appears now not only as a means of deploying efficiently the existing knowledge resources but also as strategizing for decreasing knowledge-absence and minimizing the risks associated with it in decision making. "Knowledge is about our intentional exploration of our lived context. It is not the impress of reality on our minds" (Spender 2015b, p. 63). Knowledge embraced both the objective and subjective perspectives and the etic and emic analysis of the firm demonstrates that knowledge cannot be reduced to an economical asset since it denies the scarcity assumption of tangible resources (Kahneman 2011; Nonaka and Takeuchi 1995; Spender 2015a; Zohar and Marshall 2004). Managers should go beyond the rationalist approach learned in business schools and use their experience to cope with Knight's uncertainty and Simon's bounded rationality in strategizing for the future and reducing the pressure of knowledge-absence.

2.2.2 The Knowledge-Based Theory of the Firm

We have already shown that the theories of the firm are conceptual models of real world firms and business enterprises which aim at explaining and predicting their behavior. Economic theories, as previously discussed, are concerned mainly with firm behavior in external markets. Changing the perspective, an organizational theory of the firm "analyzes the internal structure of the firm and the relationship between its constituent units and departments" (Grant 1996, p. 109).

The knowledge-based theory of the firm is an organizational theory which focuses on knowledge dynamics within the firm (Grant 1996, 1997; Nickerson and Zenger, 2004; Nonaka 1994; Nonaka and Takeuchi 1995; Spender 1996; Sveiby 2001; Tsoukas 1996). According to Grant (1997, p. 451), "the theory explains the rationale for the firm, the delineation of its boundaries, the nature of organizational capability, the distribution of decision-making authority and the determinants of strategic alliances". The key assumptions of Grant's theory are the following:

- Knowledge is the most important productive resource within a strategic perspective.
- Explicit and tacit knowledge differ in their transferability dimension. Explicit knowledge can be transferred by using a natural or symbolic language while tacit knowledge is not amenable to transfer.
- Individuals are primary agents of knowledge creation.
- Most knowledge is subject to economies of scale.

Based on these assumptions the firm is conceived as an institution able to resolve the dilemma between knowledge specialization, necessary in production, and knowledge integration through some specific mechanisms like transfer, direction, sequencing and routine. According to this logic, the principal management challenge is "establishing the mechanisms by which cooperating individuals can coordinate their activity in order to integrate their knowledge into productive activity" (Grant 1997, p. 452). Thus, managers become organizational integrators (Bratianu 2013).

Since the first assumption establishes that knowledge is the most important productive resource of the firm, the knowledge-based theory may be considered an outgrowth of the resource based view theory of the firm (Barney 1991; Barney and Clark 2007). The resource based theory conceives the firm as a bundle of specific resources and capabilities and management as a process of deploying them efficiently and effectively in achieving the firm's objectives. Introducing the strategic dimension in the managerial equation, Teece et al. (1997, p. 516) define *dynamic capabilities* as the main driving force for achieving a competitive advantage on the market: "We define dynamic capabilities as the firm's ability to integrate, build, and reconfigure internal and external competences to address rapidly changing environments. Dynamic capabilities thus reflect an organization's ability to achieve new and innovative forms of competitive advantage given path dependencies and market positions". These capabilities should not be understood in terms of balance sheets items but as managerial capacities of optimizing the allocation of resources and stimulating innovation. Since dynamic capabilities are essentially emergent functional attributes of a certain firm they cannot be replicated by competitors even if their logic can be observed by them (Teece 2009). Eisenhardt and Martin (2000, p. 1107) give a new formulation to the above definition of dynamic capabilities from a more concrete perspective. They define dynamic capabilities as: "The firm's processes that use resources—specifically the processes to integrate, reconfigure, gain and release resources—to match and even create market change. Dynamic capabilities thus are the organizational and strategic routines by which firms achieve new resource configuration as markets emerge, collide, split, evolve, and die".

The knowledge-based theory of the firm conceived by Grant (1996, 1997) identifies the following properties of *knowledge* which is the kernel of that theory: transferability, capacity for aggregation, appropriability, and specialization in knowledge acquisition. Transferability constitutes the main property of explicit knowledge and reveals the difference between explicit and tacit knowledge. Aggregation comes as property of accumulation of knowledge as any other firm's resources. However, Grant demonstrates here a linear logic of aggregation by addition, which typically applies to tangible resources but, as we have argued, does not apply to intangibles properly. For knowledge, it is better to speak of *integration* and not linear aggregation. Thus, *integration* of knowledge at individual and organizational levels is its fundamental property, and *organizational integrators* make use of that property in performing knowledge management tasks (Bratianu 2013, 2015a). In the knowledge-based theory of the firm, the whole organization is actually considered a knowledge-integrating institution, which has two main consequences: a hierarchical structure of the organization, and the distribution of decision making authority in the firm. Here, an important issue is the relationship between ownership and decision rights. Unlike tangible assets for which there is a full ownership and control by shareholders, knowledge is created and owned by people, and only when a part of it transforms into organizational knowledge through mechanisms of transfer and integration, we can speak about a partial and fuzzy ownership of it. Finally, the property of appropriability refers, in Grant's view, to "the ability of the owner of a resource to receive a return equal to the value created by that resource" (Grant 1996, p. 111). Since human capacity of storing and processing data, information and knowledge is limited, enhancing efficiency in knowledge production and use requests specialization in knowledge acquisition and processing. That is related to the fundamental assumption of the theory that knowledge represents the critical input in the production process of value creation.

A fundamental issue in the knowledge-based theory of the firm is how to define *the boundaries* of the firm (Grant 1996; Kodama 2011; Nickerson and Zenger 2004; Spender 1996, 2015b). The theory considers the intangible boundaries of the firm, which represent the span of control of its management upon the firm's resources and capabilities. This is important especially for knowledge and knowledge integration and utilization. The authors make usual reference to the vertical and horizontal boundaries with respect to other firms with which there are collaboration relationships. Knowledge boundaries relate to the boundaries defined for tangible assets, but not entirely, although "Efficient knowledge utilization requires congruence between the knowledge domain of the firm and its product domain" (Grant 1996, p. 120). An interesting analysis of knowledge boundaries and their use in creating a sustainable competitive advantage of the Japanese companies is presented by Mitsuru Kodama in his well-documented book *Knowledge integration dynamics* (2011). In Kodama's view, "Japanese companies operating in specific high-tech sectors are globally competitive because they are building their own *knowledge integration models*" (Kodama 2011, p. 37. Italics in original). These models involve the integration of internal fields of knowledge with those outside the boundaries of the firm. These fields of knowledge are generated and supported by internal and external structures (Sveiby 1997, 2001). Japanese firms excel in both vertical and horizontal knowledge integration, by following their fundamental search for harmony. We may say that they developed a successful *boundary management*. For a knowledge-based organization "Boundary management, merging, downsizing outsourcing, acquiring and so forth, in addition to changing the firm's market engagements, can precipitate energizing interpretative flexibility" (Spender 1996, p. 59).

2.2.3 The Learning Organization

By analyzing the emergence and evolution of the knowledge organization, David Bennet and Alex Bennet (2003, p. 14) consider that, thinking strategically, "A knowledge organization must, of necessity, become a learning organization so that the entire firm will learn while it works and be able to adapt quickly to market changes and other environmental perturbations". The *learning organization* is a metaphorical construct, since learning is a human process and knowledge is generated as a result of learning at the individual level. It is related to *organizational learning* and *organizational knowledge dynamics* (Argote 2013; Argyris 1999; Bratianu 2015a; Garrat 2001; Nonaka and Takeuchi 1995; Örtenblad 2011; Senge 1999; Wellman 2009).

Organizational learning can be defined as a learning process through social interactions at group and organization levels. Through organizational learning "whole organizations or their components adapt to changing environments by generating and selectively adopting organizational routines" (Argyris 1999, p. 8). Organizational learning is a generative process able to increase the level of knowledge in a certain social structure and, as a consequence, to solve a problem or to adapt to a changing environment. Organizational learning has been described by a knowledge spiral generation along the ontological dimension of a given organization by Nonaka and Takeuchi (1995), from individuals to groups and the whole organization. The spiral is evolving as knowledge is converted through the four major processes of socialization, externalization, combination, and internalization which compose the SECI dynamics model.

From a different perspective, Crossan et al. (1999) explain organizational learning as an up-ward process going through the four stages: intuinting, interpreting, integrating, and institutionalizing. *Intuinting* operates at individual level as a solving problem mechanism based on personal experience. Individual experience is examined by the unconscious zone of the brain to find out a solution to a new and rather complex problem for which little information and knowledge is available. The result of that process is called *intuition* (Gladwell 2005; Kahneman 2011; Klein 2014). The next stage of knowledge building up is *interpreting*. It operates at the interface between individual and group levels when intuition is shared within a group context. It is basically a social process when people contribute to interpreting the individual inputs and integrating them into a shared understanding. That transforms individual knowledge into group knowledge based on collective *sense-making*. Group knowledge can also be enhanced by absorbing knowledge from its social environment. That leads us to *integrating* knowledge and moving upward to the level of the organization. Integrated knowledge triggers decision making and action. Finally, the *institutionalizing* stage operates at the organization level through establishing new routines. Thus, organizational learning produces outcomes belonging to the whole organization. Although these processes have been presented as sequences in a linear logic, they have a complex and nonlinear structure.

A good example of organizational learning is *kaizen* or continuous improvement practiced by Japanese companies. For instance, Toyota developed a very successful process of continuous improvement which has been considered as a model for many other companies in the world. "Toyota works at a target condition in small, rapid steps, with learning and adjustments occurring along the way. This is the equivalent of placing one foot in front of the other, one step at a time, and always adjusting to the present situation as necessary, and is quite different than working through the predefined steps of a plan or action-item list" (Rother 2010, p. 131).

The learning organization is a result of our metaphorical thinking. As Stewart (2001, p. 147) emphasizes, "The concept of the learning organization has metaphorical status because it is embedded in the multiple narratives of organizations in all their complexity, though it becomes taken for granted, reified, and treated as though it always existed". The concept of the learning organization has been pioneered successfully by Peter Senge in his seminal book *The fifth disciplines: The art and practice of the learning organization*, published in 1990. The book became immediately a source of inspiration for many academics and practitioners to expand research and applications of this new concept related to knowledge-based organization and systems thinking. Senge considers that *the learning organization* is a social invention based on five disciplines: personal mastery, mental models, shared vision, team learning, and systems thinking. In his view, a discipline is "a body of theory and technique that must be studied and mastered to be put into practice. A discipline is a developmental path for acquiring certain skills or competencies" (Senge 1999, pp. 10–11). Personal mastery is defined as a continuous effort of learning and increasing the knowledge potential. Mental models are cognitive structures for understanding our external environment and making decisions. They are formed in schools and improved by individual experience. Shared vision means to focus on the team and organization, and to harmonize personal interests with those of the organization. Shared vision is the main enabler of strategy design. Team learning means encouraging organizational learning and creating a common platform for knowledge generation. Finally, system thinking integrates all the four disciplines and creates the necessary conditions for developing a learning organization. For Senge (1999, p. 13), "A learning organization is a place where people are continually discovering how they create their reality. And how they can change it". A learning organization should be able to go beyond the goal of surviving or adapting to a changeable environment. It should be able to achieve a competitive advantage by generating knowledge and influencing its environment. The most important message Peter Senge disseminates through his book is that we should make a shift of our mindset from focusing on single events and parts of the problem toward seeing the whole, and understanding its behavior in harmony with the external environment.

2.3 Knowledge Management

2.3.1 Knowledge Workers

Peter F. Drucker is credited with the first significant analysis of the *knowledge workers* and their *productivity* in his landmark book *The age of discontinuity*, published in 1969 and then re-published several times. With the emergence of knowledge economy, *knowledge work* and *knowledge workers* become new determinants of social value creation. Since any worker in any kind of activity uses his knowledge base which integrates his experience and learning, it is difficult to define a clear cut between industrial workers and knowledge workers. The boundary is rather fuzzy, but as we move further from it is clear that knowledge work means processing data, information and knowledge in an overwhelming way by comparison with a physical activity that is characteristic of industrial work. Although Drucker formulated no definition of knowledge workers, he identified the most important feature of their work and the new type of management necessary for that. A good working definition is offered by Davenport (2005, p. 10): "Knowledge workers have high degree of expertise, education, or experience, and the primary purpose of their jobs involves the creation, distribution, or application of knowledge".

The business of knowledge workers is thinking, creation, sharing, transfer, transformation and application of knowledge. In the knowledge economy, the activity domains where knowledge work predominates are the following: science, engineering, architecture, medicine, management, business, law, community and social services, education and training, library information, arts, design, entertainment, and media. In management, knowledge workers make decisions and find solutions to non-routine problems, and develop strategies for achieving sustainable competitive advantage on the market. In engineering, knowledge workers conceive new systems and technologies or use intelligent technologies in solving economic and social problems. In science, scientists have always been knowledge workers. What is new in the knowledge economy are the large opening of science toward society and the development of applications for people which increases the contribution of applied science and research within the whole domain of science. Following Drucker (2008, pp. 287–310) and Davenport (2005, pp. 15–22) we shall discuss some of the common attributes of knowledge workers, as follows.

Knowledge Workers Like Autonomy Since their work cannot be framed in a routine structure with well-defined and aggregated tasks within a process logic, knowledge workers have to think in their own way and thus they don't like to be told what to do and how to do it. From this point of view, managing knowledge workers is quite different than managing industrial workers where the role of managers is to define the tasks for each worker and then to control the way they perform it. The work process of knowledge workers has such a degree complexity that even for managers, it is difficult to decompose it into simpler activities and tasks and to clarify how to proceed in finding solutions. Knowledge work is intrinsically nonlinear and cannot be decomposed like a linear one. Actually, any such decomposition may change the nature of the process. Peter Senge (1999) explained this phenomenon metaphorically saying that by cutting an elephant into two pieces we don't get two smaller elephants. Nonlinearity changes completely the nature of problems and of managerial logic developed by Frederick Taylor in his scientific management. Taylor considered that each process can be decomposed in elementary units of work he called tasks, and the manager's responsibility was to document and plan how this can be done: "Perhaps the most prominent single element in modern scientific management is the task idea. The work of every workman is fully planned out by the management at least 1 day in advance, and each man receives in most cases complete written instructions, describing in detail the task which he is to accomplish, as well as the means to be used in doing the work" (Taylor 1998, p. 17). That logic is possible only for linear or quasi-linear processes. Knowledge work is a nonlinear process and cannot be decomposed following that logic. That changes completely the logic of knowledge management by comparison with classical industrial management. Relating nonlinearity with productivity in knowledge work, Drucker (2008, p. 290) states that "Two mediocre knowledge workers do not produce twice as much one first-rate one. They do not even produce as much as one mediocre knowledge worker". That is an essential fact for understanding knowledge management and how it should switch from linear to nonlinear thinking (Bratianu 2015b).

Knowledge Workers Need a Different Kind of Motivation That is an important issue since most of the managerial motivating systems are based on external rewards like money, cars, or vacations. Knowledge workers don't think only for their everyday living. They want more from their work. They want intelligent solutions and recognition from peers and managers. "What the knowledge worker needs to be positively motivated is achievement. He needs a challenge. He needs to know that he contributes. That is in complete contradiction to what we have come to consider 'good management' of the manual worker" (Drucker 2008, p. 288).

Knowledge management should be able to create new motivating and rewarding systems for stimulating knowledge workers and challenging their thinking.

Knowledge Workers Value Their Knowledge, and Don't Share It Easily Knowledge is all the knowledge workers have. It is the output of their work and the means of performing different tasks. For them, knowledge is power and a guarantee of their useful work. Sharing their knowledge, especially in a social context where competition among people is very high, may threaten their jobs. That is why knowledge workers are very reluctant to sharing their expertise for which they spent a lot of efforts and time. Knowledge management encourages people to share their knowledge and that happens currently in the Japanese firms as a result of practicing team management and of basic education focused on cooperation and team spirit (Nonaka and Takeuchi 1995).

Knowledge Worker's Productivity Means Quality Not Quantity Productivity is a key concept in classical management and is measured by means of a linear metric containing time and quantity of units produced by a worker in that given time. The concept of productivity remains important in the knowledge economy but its content should be changed. "To make knowledge work productive will be the great management task of this century, just as to make manual work productive was the great management task of the last century" (Drucker 2008, p. 290). Therefore, the measure of productivity should be changed since knowledge work is a nonlinear phenomenon. That means that quantity should be replaced by quality and a new metric should be conceived for evaluating the knowledge worker productivity. We consider that a better solution would be to create a new concept based on a nonlinear metric to measure the output of the knowledge worker and not to extend the meaning of the old concept of productivity from a linear to a nonlinear domain. That will eliminate some errors frequently made today in measuring knowledge workers' output in knowledge-intensive domains. For instance, in many European schools and universities, teachers and professors are paid according to a norm which is actually a linear metric based on physical hours of teaching. Measuring a knowledge transfer process by a linear metric is a conceptual error coming from the extension of the time norm of industrial workers to a new nonlinear domain like knowledge work. The solution of this problem can be obtained by changing completely the linear metric and not by its extension.

2.3.2 Knowledge Processes

Knowledge management cannot be properly understood if we don't identify the main knowledge processes within a firm. Let us consider a firm as an open system with well-defined boundaries of authority with respect to its environment. Then, within that firm we can identify the following basic processes: knowledge creation, knowledge acquisition, knowledge sharing and distribution, knowledge transformation, knowledge storing and retrieving, knowledge losing, and knowledge using.

Knowledge creation and knowledge acquisition contribute to the increasing level of knowledge within organization, while knowledge losing causes its decreasing. Knowledge sharing and distribution contribute to the restructuring of organizational knowledge and increase its average level. Knowledge transformation is vital for knowledge creation and decision making since it reflects the transformation of tacit into explicit knowledge on one hand, and the interaction between the rational, emotional, and spiritual knowledge fields. Knowledge storing and retrieving are functions designed for organizational memory and data bases. Knowledge losing comes into play when people with significant experience and expertise retire or leave for other jobs. Knowledge use means the convergence of all the other processes and represents the goal of knowledge management. We will briefly explain each of these knowledge processes.

Knowledge Creation One of the fundamental attribute of knowledge is that it can be created. It is created by human brain as a result of a learning process which is stimulated by body and mind interactions with natural and social environments. By integrating individuals' knowledge within a social structure and amplifying it through different processes we get organizational knowledge. Nonaka and Takeuchi (1995) explain knowledge creation as a spiral process originating in the socialization process and going through conversion of tacit knowledge into explicit knowledge, combination of explicit knowledge of employees within a Ba, conversion of explicit knowledge into tacit knowledge and starting a new cycle with knowledge sharing through socialization. We shall describe this dynamics in detail in the next chapter, but for the moment, we just wanted to show one of the most frequently used explanations of organizational knowledge creation. The process can be illustrated on a diagram with two dimensions: the epistemological and the ontological dimension. Along the epistemological dimension, the conversion from tacit into explicit knowledge, and the reverse, are represented; while along the ontological dimension, the increasing spiral from individuals to groups and organization is represented. In the paradigm of Data-Information-Knowledge-Wisdom (DIKW), knowledge creation is explained as a result of processing information which is rooted in processing raw data (Dalkir 2005; Davenport and Prusak 2000; Jashapara 2011).

Knowledge Acquisition Knowledge acquisition in organizations spans a large spectrum of activities aiming at increasing the organizational knowledge level. Knowledge acquisition refers to cognitive knowledge since emotional and spiritual knowledge are generated internally by employees. For many organizations, it can be useful to search for knowledge outside their boundaries and to acquire it in different forms. The most known methods are purchasing knowledge incorporated in data bases, scientific journals and books, or software programs. Knowledge can be acquired by hiring experts for solving some complex problems or initiating some new projects, and by hiring new people with good professional experience and records. Smart methods of knowledge acquisition can be designed as open innovation processes or by creating networks with other interested organizations in a certain business domain.

Knowledge Sharing and Distribution Knowledge sharing is a form of knowledge distribution which is distinguished from all other forms because it is based on a personal willingness to offer experience and expertise (in a certain domain or with respect to a given problem) to other people. Knowledge sharing is an unconditional process of knowledge distribution which contributes to the increase of the average level of knowledge of a team or organization. In open systems, knowledge sharing proved to be a valuable mechanism for increasing the level of organizational knowledge and organizational entropy. "Knowledge sharing has been identified as a major focus area for knowledge management. The importance of this topic lays in the fact that it aims to link the individual level, where knowledge resides, and the organizational level, where knowledge is applied and attains value" (Sanchez et al. 2013, p. 391). Knowledge sharing encounters a series of individual and organizational barriers which, aggregated together, result in knowledge *stickiness* (Szulanski 1995, 1996). A person's decision to share knowledge with somebody else depends on the trust between them. Trust is more a feeling than a rational argument, and is built on "the extent to which one is willing to ascribe good intentions to and have confidence in the words and actions of other people" (Cook and Wall 1980, p. 39). Knowledge sharing can be enhanced by creating communities of practice which are groups of people who share similar professional interests or hobbies. Unlike knowledge sharing, knowledge distribution happens as a result of managerial decisions and the needs of organization to benefit from its potential knowledge sources.

Knowledge Transformation The most known phenomena of knowledge transformation are those involved in tacit knowledge conversion into explicit knowledge through externalization, and of conversion of explicit knowledge into tacit knowledge through internalization. Both the processes of externalization and internalization are contained in the cyclic SECI model initiated by Nonaka (1994) and then developed further by Nonaka and his colleagues (Nonaka and Takeuchi 1995; Nonaka et al. 2008). As we presented in Chap. 1, the new theory of knowledge dynamics developed by Bratianu (2011, 2015a, b), based on metaphorical analysis, introduces the multi-field theory of knowledge where three fundamental fields of knowledge are defined: rational, emotional, and spiritual. There is a continuous interaction between these fields, and transformations of one form of knowledge into another. This knowledge dynamics is important in understanding the decision making process and the performance of the firm.

Knowledge Storing and Retrieving These processes are possible at both individual and organizational levels. Employees have the capacity of memorizing data, information and knowledge and retrieving it whenever it is necessary. At the organizational level, there are databases created and supported by the information technology where data, information and knowledge can be stored in some codified structures in order to be easily retrieved. Also, knowledge can be embedded in different routines, regulations, patents and organizational culture which together can be identified as the organizational memory.

Knowledge Loss When people retire, they take their experience and expertise with them, which represents a loss for the firm. When a large group of employees retire at the same time, the loss of knowledge becomes significant and the firm may lose its competitive advantage (DeLong 2004). For instance, the retirement wave generated by baby boomers will lead to substantial knowledge loss if knowledge management does not create some strategies to reduce it. Also, knowledge loss can become a problem when there is a downsizing or reengineering of organizations. NASA could be such a negative example due to its many restructuring processes and massive retirement phenomena (Mahler and Casamayou 2009).

Knowledge Use This is the final stage of any knowledge process: embedding it in goods and services. Smart products contain a high level of knowledge and intelligence embedded in their design, production and marketing. For instance, smartphones integrate a huge amount and a high quality of knowledge in their software, which make them very expensive products by comparison with their manufacturing costs. Knowledge use constitutes the kernel of knowledge management and the efficiency and productivity metrics of classical management have to be changed with new metrics able to measure the density of embedded knowledge and intelligence in goods and services.

2.3.3 Knowledge Management as an Emergent Process

In this Chapter, we have chosen an evolutionary approach in order to build up the conceptual framework of knowledge management and to show how it is a product of the evolving dynamics of all the changes happened in the last 50 years in economy and organizational management. We decided to avoid starting with definitions, more or less adequate to the nature of knowledge management, in order to have an open view and a critical thinking approach rather than a prescriptive framework. Knowledge management is a new process but its emergence needed some historical time and significant changes in the post-industrial society. Knowledge management is not a fad or a new managerial fashion with a short life cycle. It is a different process since it involves intangible resources which are completely different from the tangible ones, and intangible processes which replace assembly lines and mass production. The "magical" economic concepts of *efficiency* and *productivity* lost their primary meanings and all linear metrics used so far for measuring production inputs and outputs proved to be outdated. Instead of enabling knowledge management, they constrain it by offering false interpretations of final results. Knowledge management is not an extrapolation of information management, although many authors even today use these thinking lenses, which are specific for it and for a technological approach.

The emergence of knowledge economy and knowledge-based organizations created the necessary conditions for the emergence of *knowledge management*, as a new managerial process able to deal with knowledge workers and knowledge processes. Knowledge management does not replace the classical management, but brings a new type of processes within the framework of organizational management: the knowledge-based processes. These processes are distinguished from other industrial processes due to operations with intangible resources. That means resources that cannot be seen, cannot be touched and as a consequence cannot be measured in the same way like tangible ones.

Knowledge management emerged as a new process in the organizational practice and at the same time as a new research domain for academics. Inputs and contributions for developing knowledge management came from a large variety of people with different professional backgrounds and professional philosophies, ranging from management to information sciences, and from epistemology to psychology and sociology. Also, it is interesting to remark that they came from different cultures which means different mindsets for understanding and working with knowledge. The explanation of such a large variety comes from the fact that *knowledge* is a universal concept but it is defined through the lenses of each specific domain. Also, managers and academics have different approaches to working with knowledge since, for managers, tacit knowledge plays the dominant role, while for academics, explicit knowledge does it. Moreover, for managers educated in the European and American universities, rational knowledge constitutes the basis of any analysis and decision making, while for managers educated in Japanese, Korean or Chinese universities, there is a balance between rational, emotional, and spiritual fields of knowledge. That means that knowledge management will emphasize primarily the analytics, mathematical decision making, and competition in Western organizations, while in the Eastern organizations the emphasis will be on knowledge sharing, judgment, and cooperation (Davenport and Harris 2007; Nonaka and Zhu 2012; Spender 2014).

Many authors, coming from information management, consider knowledge management just an extension of it, and overemphasize the role of information technology and Big Data in knowledge processes. It is true that information technology proved to be an efficient enabler of knowledge management, but it cannot replace the human dimensions. People trained in information science and information management use the concept of *information* in a different metric than those trained in social sciences. In the first case, information is just a mathematical concept which reflects a probability distribution of a data set, as it has been defined by Claude E. Shannon (see Chap. 1). Even though experts in computer science discuss about semantic information and semantic webs, they still use mathematical methods of analysis and cannot reach to knowledge as we discussed it in the previous chapter. In knowledge management, we use the Data-Information-Knowledge-Wisdom (DIKW) paradigm which gives a different content to information (Davenport and Prusak 2000; Dalkir 2005). In this new framework, *information* is a result of processing data and emerging in a field of meanings. Its importance increases when we consider that, by processing information, we get knowledge.

People who are not aware of this distinction may generate confusion when they express their views about knowledge management.

Hislop (2005), in a critical analysis of knowledge management based on the research done by authors like Alvesson and Karreman (2001) and Scarbrough and Swan (2001), presented some weak points and inconsistencies of its theoretical foundations. The main aspects of these criticisms focused on the fact that basic concepts like "knowledge", "management", and "knowledge management" have fuzzy definitions and, sometimes, even contradictory formulations. That generated some ambiguity in understanding knowledge management and in implementing it adequately. We have shown in the first chapter that understanding knowledge is a metaphorical approach, and its interpretation depends on the kind of metaphors people use. Thus, it is explainable why different authors may get different meanings of knowledge and, as a consequence, of knowledge management.

The main limitations in understanding, explaining and especially implementing knowledge management come from the fact that it operates with resources which have two fundamental attributes that are completely different from physical resources: nonlinearity and intangibility. Nonlinearity creates real difficulties to all managers and academics who want to measure and process quantitatively knowledge by using linear metrics and mathematical models. Since these measurements cannot justify the managerial practice in organizations, people blame knowledge management for this. Indeed, we should understand that knowledge management needs a new paradigm to explain complexity and nonlinearity of intangible assets, and not to force it into the frame of the linear paradigm of industrial management. We have already discussed the new problems of motivating knowledge workers and measuring the outcomes of their work, which cannot be judged in terms of efficiency and productivity of the economics of tangible resources. The intangibility of knowledge resources and processes should lead to a new conception of management, since managers cannot operate and control them directly. Von Krogh et al. (2000, p. 17) state this fact very clearly: "The real point is that while you may be able to manage related organizational processes like community building and knowledge exchange, you cannot manage knowledge itself. Those who try to control knowledge creation do so at their peril, often putting up barriers or falling into the pitfalls".

Knowledge workers need autonomy in planning and organizing their work which means that the difference between managers and knowledge workers decreased such that managers lost some of their authority. Based on this phenomenon, Davenport (2005) shows that as industrial workers created a need for professional managers able to plan, organize, lead and control their work, the emergence of a critical mass of knowledge workers now calls for a new managerial approach. "Because knowledge work can be and is done by managers as well as workers, strict separations between worker and manager no longer make sense. Because knowledge work has become the key to growth and differentiation in today's economy, the differential in cost and value between knowledge work and management has decreased. Management in the 'knowledge economy' is a different game, with different rules" (Davenport 2005, p. 191).

Since knowledge is considered a strategic resource in achieving competitive advantage by many experts, knowledge management has a clear strategic dimension. In other words, knowledge management bridges the gap between operational management and strategic management, leading to strategic thinking and strategies design. However, we should go beyond seeing knowledge only as a resource, since knowledge is also vital in decision making. That means to change the meaning of *productivity* applied to knowledge workers or to define a new metric able to show that, for knowledge management, what matters is the quality of decision making and not the quantity of knowledge or ideas issued per unit time. The new mantra should be *working smarter* instead of *working harder*.

2.4 Conclusion

Knowledge economy is a fact today. Anticipated by Peter Drucker since 1969 in his famous book about economic discontinuities, the knowledge economy dominates in developed countries like United States, Japan, Korea, Northern and Western European countries. It is a new economy based primarily on intangible resources and on knowledge processes. From the scarcity of physical resources, economic theories of profit maximization and linear metrics for measuring efficiency and workers' productivity, things now changed to an abundance of intangible resources, economic theories of sustainability, and nonlinear metrics for intangibles which have not been defined yet. According to the World Bank Institute (2008) the major pillars of the new economy are the following: economic and institutional regime, education and skill of population, information infrastructure, and innovation system. Institutional regime should promote new legislation and administrative structures which stimulate knowledge creation, dissemination, and use as well as intellectual property. Education is important not only in quantitative terms of people graduating schools and universities but mostly in terms of developing generic skills necessary for new knowledge processes. Information infrastructure supports knowledge processes and enables knowledge workers to share their ideas and research findings with other people without any space and time constraints. Renewing the intellectual capital of organizations and achieving competitive advantage on a global market requires creativity and developing innovation systems. As demonstrated by Richard Florida (2002), with knowledge economy a new class of people is restructuring the society—the creative class.

The engine of knowledge economy is the knowledge-based organization. As JC Spender underlined in most of his papers and books, the merit of analyzing the nature of the firm and its driving forces for growth must be ascribed to Edith Penrose, who explored the role of knowledge and their attributes as critical resources for any organization. Also, she showed how decision making is always subject to uncertainty and bounded rationality, as demonstrated by Herbert Simon (1976). Putting together these ideas, we get the image of the knowledge-based organization as being primarily a bundle of intangible resources and dynamic capabilities that promote organizational learning and aiming at achieving a sustainable competitive advantage. When organizational knowledge processes reach the whole organizational structure, the firm becomes a learning organization. According to Peter Senge (1999), the main purpose of such a learning organization is to generate knowledge not only for its adaptation to a changeable environment but to contribute to its transformation. The learning organization is not a state-of-the-art but an ideal configuration to which any knowledge-based organization should converge. It is like a strange attractor in the theory of complexity.

In knowledge-based organizations, the critical mass of employees are knowledge workers. Their business is thinking and knowledge creation. They are not aggregated by assembly lines but integrated in networks and communities of professionals. Their work is not based on a physical effort but on an intellectual one, and their work schedule is not synchronized with a whole shift but desynchronized and made almost independent. Information technology shrinks space and time and allows global collaboration. Knowledge assets are owned by employees and can be only partially controlled by managers, in concordance with job requirements. Knowledge workers are less motivated by external rewards and more attracted by internal ones like job satisfaction and relevance of their contribution to the advance of knowledge in a certain field. Knowledge, being an intangible resource, cannot be managed directly as a physical resource but only through knowledge workers within a stimulating environment. Knowledge management has to optimize organizational knowledge dynamics and to motivate knowledge workers in powering business innovation.

Knowledge management emerged as a necessary process to deal with knowledge as the critical resource of organization, knowledge workers and knowledge work. Knowledge management is integrated in the firm's management but it departs from industrial management since it operates with intangible resources. Knowledge management bridges the gap between operational management and strategic management, since knowledge represents strategic resources playing a crucial role in firm's competitive advantage. For knowledge management, strategic thinking becomes vital in designing strategies able to cope with future uncertainty and bounded rationality.

References

Al-Ali, N. (2003). *Comprehensive intellectual capital management*. New York: Wiley.

Alvesson, M., & Karreman, D. (2001). Odd couple: Making sense of the curious concept of knowledge management. *Journal of Management Studies, 38*(7), 995–1018.

Argote, L. (2013). *Organizational learning: Creating, retaining and transferring knowledge* (2nd ed.). Heidelberg: Springer.

Argyris, C. (1999). *On organizational learning* (2nd ed.). Oxford: Blackwell.

Barney, J. B. (1991). Firm resources and sustained competitive advantage. *Journal of Management, 17*(1), 99–120.

Barney, J. B., & Clark, D. N. (2007). *Resource-based theory: Creating and sustaining competitive advantage*. Oxford: Oxford University Press.

Bell, D. (1999). *The coming of post-industrial society: A venture in social forecasting*. New York: Basic Books.

Bennet, D., & Bennet, A. (2003). The rise of the knowledge organization. In C. W. Holsapple (Ed.), *Handbook on knowledge management* (Vol. 1, pp. 5–21). Heidelberg: Springer.
Bratianu, C. (2009). The frontier of linearity in the intellectual capital metaphor. *Electronic Journal of Knowledge Management, 7*(4), 415–424.
Bratianu, C. (2011). Changing paradigm for knowledge metaphors from dynamics to thermodynamics. *Systems Research and Behavioral Science, 28*(2), 160–169.
Bratianu, C. (2013). Nonlinear integrators of the organizational intellectual capital. In M. Fathi (Ed.), *Integration of practice-oriented knowledge technology: Trends and perspectives* (pp. 3–16). Heidelberg: Springer.
Bratianu, C. (2015a). *Organizational knowledge dynamics: Managing knowledge creation, acquisition, sharing, and transformation*. Hershey: IGI Global.
Bratianu, C. (2015b). Developing strategic thinking in business education. *Management Dynamics in the Knowledge Economy, 3*(3), 409–429.
Coase, R. H. (1937). The nature of the firm. *Economica, 4*(16), 386–405.
Cook, J., & Wall, T. (1980). New work attitude measure of trust, organizational commitment and personal need nonfulfillment. *Journal of Occupational Psychology, 53*(1), 39–52.
Crossan, M. M., Lane, H. W., & White, R. E. (1999). An organizational learning framework: From intuition to institution. *Academy of Management Review, 24*(3), 532–537.
Dalkir, K. (2005). *Knowledge management in theory and practice*. Amsterdam: Elsevier.
Davenport, T. H. (2005). *Thinking for a living: How to get better performance and results from knowledge workers*. Boston: Harvard Business School Press.
Davenport, T. H., & Harris, J. G. (2007). *Competing of analytics: The new science of winning*. Boston: Harvard Business School Press.
Davenport, T. H., & Prusak, L. (2000). *Working knowledge: How organizations manage what they know*. Boston: Harvard Business School Press.
DeLong, D. W. (2004). *Lost knowledge: Confronting the threat of an aging workforce*. Oxford: Oxford University Press.
Drucker, P. F. (2008). *The age of discontinuity: Guidelines to our changing society*. New Brunswick: Transaction Publishers.
Eisenhardt, K. M., & Martin, J. A. (2000). Dynamic capabilities: What they are? *Strategic Management Journal, 21*(10/11), 1105–1121.
Florida, R. (2002). *The rise of the creative class: And how it's transforming work, leisure, community and everyday life*. New York: Basic Books.
Garrat, B. (2001). *The learning organization: Developing democracy at work*. London: Harper Collins Business.
Gladwell, M. (2005). *Blink: The power of thinking without thinking*. New York: Back Bay Books.
Grant, R. M. (1996). Toward a knowledge-based theory of the firm. *Strategic Management Journal, 17*(Winter), 109–122.
Grant, R. M. (1997). The knowledge-based view of the firm: Implications for management practice. *Long Range Planning, 30*(3), 450–454.
Hislop, D. (2005). *Knowledge management in organizations. A critical introduction*. Oxford: Oxford University Press.
Jashapara, A. (2011). *Knowledge management: An integrated approach* (2nd ed.). London: Financial Times/Prentice Hall.
Kahneman, D. (2011). *Thinking, fast and slow*. New York: Farrar, Straus and Giroux.
Klein, G. (2014). *The power of intuition: How to use your gut feelings to make better decisions at work*. New York: Currency Doubleday.
Knight, F. H. (1921). *Risk, uncertainty and profit*. Boston: Houghton Mifflin.
Kodama, M. (2011). *Knowledge integration dynamic: Developing strategic innovation capability*. Singapore: World Scientific.
Leydesdorff, L. (2006). The knowledge-based economy and the triple helix model. In W. Dolfsma & L. Soete (Eds.), *Understanding the dynamics of a knowledge economy* (pp. 42–76). Cheltenham: Edward Elgar.

Machlup, F. (1962). *Production and distribution of knowledge in the United States*. Princeton: Princeton University Press.
Mahler, J. G., & Casamayou, M. H. (2009). *Organizational learning at NASA: The challenger and Columbia accidents*. Washington, D.C.: Georgetown University Press.
Nickerson, J. A., & Zenger, T. R. (2004). A knowledge-based theory of the firm: The problem-solving perspective. *Organization Science, 15*(6), 617–632.
Nonaka, I. (1994). A dynamic theory of organizational knowledge creation. *Organization Science, 5*(1), 14–37.
Nonaka, I., & Takeuchi, H. (1995). *The knowledge-creating company: How Japanese companies create the dynamics of innovation*. New York: Oxford University Press.
Nonaka, I., & Zhu, Z. (2012). *Pragmatic strategy: Eastern wisdom, global success*. Cambridge: Cambridge University Press.
Nonaka, I., Toyama, R., & Hirata, T. (2008). *Managing flow: A process theory of the knowledge-based firm*. Houndmills: Palgrave Macmillan.
North, K., & Gueldenberg, S. (2011). *Effective knowledge work: Answers to the management challenges of the 21st century*. Bingley: Emerald.
O'Dell, C., & Hubert, C. (2011). *The new edge in knowledge: How knowledge management is changing the way we do business*. New York: Wiley.
Örtenblad, A. (2011). *Making sense of the learning organization: What is it and who needs it?* Kuala Lumpur: YayasanIlmuwan.
Penrose, E. T. (2013). *The theory of the growth of the firm*. Mansfield Centre: Martino Publishing.
Powell, W. W., & Snellman, K. (2004). The knowledge economy. *Annual Review of Sociology, 30*, 199–220.
Rother, M. (2010). *Toyota kata: Managing people for improvement, adaptiveness, and superior results*. New York: McGrawHill.
Ryle, G. (1949). *The concept of mind*. Chicago: The University of Chicago Press.
Sanchez, J. H., Sanchez, Y. H., Collado-Tuiz, D., & Cebrian-Tarrason, D. (2013). Knowledge creating and sharing corporate culture framework. *Procedia: Social and Behavioral Sciences, 74*, 388–397.
Scarbrough, H., & Swan, J. (2001). Explaining the diffusion of knowledge management. *British Journal of Management, 12*, 3–12.
Senge, P. (1999). *The fifth discipline: The art and practice of the learning organization*. London: Random house.
Simon, H. A. (1976). *Administrative behavior* (3rd ed.). New York: Macmillan.
Simon, H. A. (1991). Bounded rationality and organizational learning. *Organization Science, 2*, 125–134.
Spender, J. C. (1996). Making knowledge the basis of a dynamic theory of the firm. *Strategic Management Journal, 17*(Winter), 45–62.
Spender, J. C. (2014). *Business strategy: Managing uncertainty, opportunity, and enterprise*. Oxford: Oxford University Press.
Spender, J. C. (2015a). Knowledge management: Origins, history, and development. In E. Bolisani & M. Handzic (Eds.), *Advances in knowledge management: Celebrating twenty years of research and practice* (pp. 3–25). Heidelberg: Springer.
Spender, J. C. (2015b). The theory of the managed firm (TMF). *Human Systems Management, 34* (1), 57–80.
Stewart, D. (2001). Reinterpreting the learning organization. *The Learning Organization, 8*(4), 141–152.
Sveiby, K. E. (1997). *The new organizational wealth – managing and measuring knowledge-based assets*. San Francisco: Berrett-Koehler.
Sveiby, K. E. (2001). A knowledge-based theory of the firm to guide in strategy formulation. *Journal of Intellectual Capital, 2*(4), 344–358.
Szulanski, G. (1995). Unpacking stickiness: An empirical investigation of the barriers to transfer best practices inside the firm. *Academy of Management Proceedings, 38*, 437–441.

Szulanski, G. (1996). Exploring internal stickiness: Impediments to the transfer of best practice within the firm. *Strategic Management Journal, 17*(Winter), 27–43.
Taylor, F. W. (1998). *The principles of scientific management*. New York: Dover Publications.
Teece, D. J. (2009). *Dynamic capabilities and strategic management. Organizing for innovation and growth*. Oxford: Oxford University Press.
Teece, D. J., Pissano, G., & Shuen, A. (1997). Dynamic capabilities and strategic management. *Strategic Management Journal, 18*(7), 509–533.
Tsoukas, H. (1996). The firm as a distributed knowledge system: A constructionist approach. *Strategic Management Journal, 17*(Winter), 11–25.
Von Krogh, G., Ichijo, K., & Nonaka, I. (2000). *Enabling knowledge creation: How to unlock the mystery of tacit knowledge and release the power of information*. Oxford: Oxford University Press.
Wellman, J. L. (2009). *Organizational learning: How companies and institutions manage and apply knowledge*. New York: Palgrave Macmillan.
World Bank Institute. (2008). *Measuring knowledge in the world's economies: Knowledge assessment methodology and knowledge economy index*. Washington, DC: World Bank.
Zohar, D., & Marshall, I. (2004). *Spiritual capital: Wealth we can live by*. San Francisco: Berret-Koehler Publishers.

Knowledge as a Strategic Weapon

3

Knowledge has long been accepted as a strategic asset to achieve and maintain competitive advantage. The purpose of this chapter is to show how, in the literature, the connection between knowledge and strategy has been more or less clearly underlined. In the first part of this chapter, we will show why knowledge is increasingly considered one of the fundamental elements of value creation in business. Its strategic attribute became essential especially in the last decades, due to the turbulence of the business environment. Since the creation of economic value is the most important requirement for achieving success, this soon brings us to the idea of a strategy as a means of realizing it. In Sect. 3.2, a discussion about the notion of strategy will be provided, starting with the military original meanings of it. In Sect. 3.3 we will show how this notion has evolved over time as a result of a new equilibrium of forces between internal and external business environment. Finally, we will illustrate that knowledge has always been a central element in all the various perspectives on strategy and strategic thinking that can be found in the literature and managerial practice.

3.1 Strategic Knowledge and Value Creation

3.1.1 Knowledge as a Strategic Resource and Capability

As we recalled in the previous chapter, in the Resource-Based View (RBV) managerial philosophy, the basic assumption is that the firm's performance depends on the quality and quantity of resources and capabilities owned and controlled by the firm (Barney 1991; Barney and Hesterly 2012). However, resources and capabilities are not homogeneous and they do not make the same contribution to value creation. Similarly, not all knowledge has strategic value. In particular, according to RVB, strategic resources should be able to pass the VRIO test: "Valuable, rare, and costly-to-imitate resources and capabilities can be a source of sustained competitive advantage. Imitation can occur through direct duplication

E. Bolisani, C. Bratianu, *Emergent Knowledge Strategies*, Knowledge Management and Organizational Learning 4, DOI 10.1007/978-3-319-60657-6_3

or through substitution. A firm's resources and capabilities may be costly to imitate for at least four reasons: unique historical circumstances, causal ambiguity, socially complex resources and capabilities, and patents" (Barney and Hesterly 2012, p. 113).

RBV has been taken as a basis for developing *the knowledge-based view of the firm* (Grant 1996, 1997; Nonaka 1994; Spender 1996; Sveiby 2001). In the knowledge economy, knowledge is a valuable, rare and costly-to-imitate resource when used for new products and services. Knowledge is an intangible resource and, especially when it is in a tacit form, it is difficult to imitate it. Spender (1996, p. 46) remarks that specific aspect very well: "Since the origin of all tangible resources lies outside the firm, it follows that competitive advantage is more likely to arise from the intangible firm-specific knowledge which enables it to add value to the incoming factors of production in a relatively unique manner". Knowledge is the raw material of developing capabilities of a firm. *A capability* represents a capacity for a team or an organization to use efficiently its resources. That includes also the knowledge about *how to do it*. If knowledge as a resource is interpreted as a result of knowing *what*, then a capability is knowledge embodied in knowing *how* to perform a certain task. Thus, knowledge is the hidden component of organizational routines, procedures and managerial thinking. Knowledge is also the fundamental ingredient of organizational culture and working spirituality, especially under the forms of emotional and spiritual knowledge (Bratianu 2015). For instance, "3M's capabilities in developing new product and fostering their market and commercial development rests upon a set of routines, practices and attitudes that are built into the company's culture as well as into its operating procedure" (Grant 1997, p. 452).

Another illustration of this fundamental aspect is given by the *Toyota system* of production: "Toyota's tools and techniques, the things you see, are built upon invisible routines of thinking and acting, particularly in management, that differ significantly from those found in most companies" (Rother 2010, p. 4). The automobile companies, especially in the United States, that analyzed Toyota's market success, discovered lean techniques such as *kanban*, cellular manufacturing, short changeovers, and reverse engineering processes, that contributed to achieving its well-known competitive advantage. But when they tried to imitate Toyota and to implement these techniques in their production systems, the results were far lower from those anticipated. The explanation comes from the fact that, in each technology or routine, there is deep knowledge embedded which integrates rational, emotional, and spiritual knowledge. While rational knowledge is more easily imitable, emotional and spiritual knowledge cannot be imitated and replicated. Emotional and spiritual knowledge is culture-specific and context-embodied. For instance, when Toyota elaborated the new international strategies aiming at a significant geographic expansion, the emotional and spiritual knowledge became crucial: "Toyota has also realized an urgent need to clearly articulate and implement a set of common values, beliefs, and business methods, some of which are tacit and shared face-to-face, to support and guide the continuing evolution of its global operations" (Ichijo 2007, p. 129). That explains the hidden part of the competitive advantage of the Japanese car makers with respect to the American companies.

Nonaka (1994) considers that a strategic capability on which Japanese companies focused in the last decades is *organizational knowledge creation.* As shown in the first two chapters of this book, knowledge is created by individuals and then amplified through collaboration to the level of teams and then to the organizational level. *Organizational knowledge* is a semantic construct which reflects the capacity of an organization to integrate all the employees' knowledge contributions and to generate a collective knowledge. Referring to this process within the knowledge organizations, Brown and Duguit (1998, p. 91) state that "Their ability to outperform the marketplace rests on the continuous generation and synthesis of collective, organizational knowledge. For all organizations, the cultivation of this knowledge—often an implicit, unreflecting cultivation—is the essence of developing a core competency to maintain organization and resist its dissolution". Nonaka (1994) conceived a dynamic model of creating organizational knowledge composed of four basic conversions: socialization—tacit knowledge from one individual is transferred to tacit knowledge of another individual; externalization—tacit knowledge is transformed into explicit knowledge at individual level; combination—explicit knowledge from one individual is combined with explicit knowledge of other individuals; internalization—explicit knowledge is transformed into tacit knowledge at individual level. This is a brief description of the well-known SECI model that we also mentioned in the previous chapters. The model generates a knowledge spiral which reflects knowledge creation as a continuous iteration between tacit and explicit knowledge along the epistemological dimension. Nonaka and Takeuchi (1995, p. viii) consider that it is exactly this spiral development of organizational knowledge creation that contributed to achieving a competitive advantage by Japanese companies: "We contend in this book that Japanese companies have become successful because of their skills and expertise at 'organizational knowledge creation'. By organizational knowledge creation we mean the capability of a company as a whole to create new knowledge, disseminate it throughout the organization, and embody it in products, services, and systems".

Organizational knowledge creation is closely linked to another strategic capability which is *organizational learning.* It is a strategic capability since knowledge is a product of the learning process, and organizational knowledge is a result of the organizational learning process. Japanese companies developed this strategic capability starting from a new perspective of organizational learning defined by the insight that failures can be as useful as successes in extracting critical knowledge. For instance, "Toyota's approach of not stigmatizing failures, but instead utilizing them to learn and move forward, has an interesting effect: thinking that an abnormality or problem is neither positive nor negative shifts the focus from individual to the process. We know that the vast majority of problems are caused by the system within which people work, rather than by the individuals themselves" (Rother 2010, p. 141).

Knowledge creation and organizational learning are two powerful driving forces of *innovation*, which is a complete process of *knowledge intensive business* starting with exploitation of the available knowledge, and continuing with exploring and creating new knowledge, embedding that knowledge into new products and services, and selling them on the market. Pasher and Ronen (2011, p. 166) consider that "In today's knowledge economy, if you don't innovate, you die. The strategic race is about finding and utilizing unorthodox ideas that create the future by changing the rules of an industry; it's about redrawing boundaries between industries, and creating entirely new industries". Innovation is about generating new business ideas and transforming them into successful products and services. That means integrating new knowledge with existing knowledge, reshaping its structure, and changing its destination. Augier and Teece (2007, p. 200) define innovation as "the search for and the discovery, development, improvement, adaptation, and commercialization of new processes, new products, and new organizational structures and procedures". Innovation is advancing in the world of uncertainty with new ideas and taking risks of possible failure. If managers are risk-averse they will choose an *incremental innovation* strategy, which means a process of small improvements to already existing products and services. If managers are risk-seekers and visionary in their business domain, then they will take the opportunity of *disruptive technology* to create *disruptive innovations* (Christensen 2003). These innovations result in totally new products and services based on new and powerful technologies. However, in the beginning, these new products cannot compete on quality and price with the similar old ones, but, as time goes on, they will be improved and will probably replace the other products. Digital cameras can be a good example. Initially, digital photos could not have the quality of those produced by Kodak cameras, reason for which Kodak ignored the digital technology completely. In time, due to many technology improvements, digital cameras produced comparable and then even better images than old cameras with chemical films. That disruptive innovation created an incredible competitive advantage for companies producing digital cameras, and eliminated the old type of cameras completely from the market.

Today, strategic use of knowledge as a resource and production capability may lead to the creation of exponential technologies and *Big Bang Disruptions*. According to the research performed by Downes and Nunes (2014, p. 7), "We have entered a fourth stage of innovation—the era of Big Bang Disruption. The new disrupters attack existing markets not just from the top, bottom, and sides, but from all three at once. By tying their products to the exponential growth and falling costs of new technologies, their offerings can be simultaneously better, cheaper, and more customized. Not just for one group of users, but for all (or nearly all) customers. This isn't disruptive innovation. It's devastating innovation". That is, metaphorically speaking, a nuclear knowledge weapon able to devastate a market and to change completely the direction of business. For instance, the innovation of the smartphone changes the idea of mobile phones completely, from a simple telephone technology into a complex integration of telephone, internet and computing technologies. That means that big disrupters change also the way we think about business and knowledge strategies, since there is also a change in the business rules. Big disrupters can be considered as *black swans* in the new business landscape in the perspective described by Taleb (2008). These are highly improbable events which hardly can be anticipated. As Taleb (2008, p. xix) explains, "Black Swan logic makes *what you don't know* far more relevant than what you do know. Consider that many Black Swans can be caused and exacerbated *by their being unexpected*".

3.1.2 Knowledge Value Chain

As affirmed by Holsapple and Singh (2003, p. 216), citing Stewart (1997), "Knowledge is the thermonuclear competitive weapon of our time; knowledge and its management are more valuable and more powerful than natural resources, big factories, or fat bankrolls". Maybe it is not thermonuclear, but for sure it can be seen as a weapon in business, because it is deemed to allow companies to improve their capability to make valuable profit. The recalled upsurge of the Knowledge economy or Knowledge-based economy is associated with an idea of knowledge as an input and output of economic activities (Prusak 2009). Indeed, knowledge is an ingredient or, better, a production factor that every company uses, explicitly or not. But now, "knowledge, it seems, has become the most important or 'strategic' factor of production, so managers must now focus on its production, acquisition, movement, retention and application" (Spender 1996, p. 48). In addition, for some companies, knowledge is their deliberate output, i.e. what they sell to customers: entire categories of firms (media companies, consulting societies, business services, etc.) sell knowledge incorporated in their products or services. For example, an entire economic sector has been newly associated with the idea of *knowledge* as main economic element: the so-called *Knowledge Intensive Business Services (KIBS)*. The functioning of KIBS is strictly associated with cognitive assets that they not only acquire or employ, but also sell and exchange with customers. If knowledge becomes a production factor or an outcome of economic activity (just like products, materials, money, labour), this explains why it is important to include it into any reasoning about a strategy for creating value in business.

Also, knowledge becomes important to understand how companies produce value internally. A popular scheme that models value creation in companies is Porter's (1985) value chain, where fundamental activities in a company are singled out to understand their individual contribution to the creation of economic value. It is interesting to notice that the same has been done as regards knowledge and KM: the so-called "knowledge chain". This model "posits nine distinct, generic activities that an organization performs in the course of managing its knowledge resources. These are focal points for achieving competitiveness through knowledge management" (Holsapple and Singh 2003, p. 217). In short, it is underlined that each KM activity can bring value to the business. The *Knowledge Chain Model* contains five primary activities which contribute directly to value creation and four secondary or supportive activities. According to Holsapple and Singh (2003, p. 221), the primary activities of the value chain are the following:

- Knowledge acquisition—Obtaining knowledge from the external business environment and adapting it to the needs of the organization. There are different ways of doing knowledge acquisition which will be discussed when presenting the knowledge generic strategies, in a future chapter.
- Knowledge selection—Selecting needed knowledge from all possible internal sources and preparing it for suitable use in the value chain.
- Knowledge generation—Knowledge creation at individual level and then amplifying it at the organizational level.
- Knowledge internalization—Distributing, sharing, transforming and storing in information and knowledge bases organizational knowledge.
- Knowledge externalization—Embedding knowledge into organizational outputs for release into the external business environment.

These primary activities have the support of four secondary activities:

- Knowledge leadership—That is the most important activity from strategic point of view. Without leadership there is no vision and no long term managerial thinking.
- Knowledge coordination—That is a part of operational management which should integrate all knowledge activities in organization.
- Knowledge control—Ensuring that needed knowledge processors and resources are available as requested, both from quality and quantity viewpoints.
- Knowledge measurement—Evaluating as much as it can be done knowledge resources, capabilities, processes and their deployment.

The Knowledge Chain Model evidences the key activities and processes involving knowledge management, and shows the focal aspects of the organizational knowledge dynamics in a strategic perspective. However, we should remark that Holsapple and Singh have used the term "activities" in a generic sense because they are complex processes composed of a number of activities. For instance, "knowledge leadership" represents an integrated component of the organizational leadership which cannot be reduced to a simple activity: "The distinguishing characteristics of leadership is that of being a catalyst through such traits as inspiring, mentoring, setting examples, engendering trust and respect, instilling a cohesive and creative culture, establishing a vision, listening, learning, teaching, and knowledge sharing" (Holsapple and Singh 2003, p. 244). Thus, the model provides a conceptual framework for knowledge value chain and its role in achieving competitiveness, rather than identifying all knowledge management activities in detail. As a consequence of its concrete application within a knowledge-based organization, top management should be able to create and implement knowledge strategies as practical ways of achieving competitive advantage.

One of the most difficult decision making area is finding the best balance between *outsourcing* and *insourcing* of knowledge, by making an analogy and taking advantage of the best practices in manufacturing (Jenster et al. 2005). For instance, fast-growing off-shoring activities for KIBS are mostly call centers, designing and testing different information technologies, and even R&D activities. As North and Gueldenberg (2011, p. 159) remark, "From the point of view of emerging economies, off-shoring creates many opportunities, for instance, a higher level of education, economic growth and as a result a generally higher standard of living". The performance of new information technologies and their powerful applications creates many opportunities for outsourcing knowledge-based activities, although this topic is rather debatable (see e.g. the famous "productivity paradox"—Triplett 1999). But for sure, what is clear is that the ubiquitous presence of interconnected devices to collect, store and process data is challenging not only developers of software applications, but also analysts. It is clear that these huge amounts of data can be strategic for competitive intelligence (Minelli et al. 2012), but what is really arduous is how to transform them into useful knowledge (Lamont 2012; Kabir and Carayannis 2013). The connection between big data and KM still needs to be explored, but, for sure, it will be a key issue in the future management studies. This brief outline is only a selection of some crucial points that can be found in the literature. However, it is sufficient for understanding that knowledge value creation is not a mere abstract statement, but has important implications not only for research but also for the managerial practice.

3.1.3 Strategic Knowledge and Intellectual Capital

The new managerial paradigm that describes knowledge as *a strategic resource and capability* reflects, like in a mirror, the creation of the economic paradigm of *intellectual capital.* The common essence of these two paradigms has been appropriately anticipated by Peter Drucker who is the father of the expression "knowledge economy" (Drucker 1969). In a more recent book, Drucker (1993, p. 42) remarks that knowledge is the only meaningful resource today and that "The traditional 'factors of production'—land (i.e. natural resources), labor and capital—have not disappeared, but they have become secondary. They can be obtained, and obtained easily, provided there is knowledge. And knowledge in this new sense means knowledge as a utility, knowledge as the means to obtain social and economic results". This last idea leads easily to the concept of intellectual capital defined by Stewart (1997, p. xi) as being "knowledge, information, intellectual property, experience—that can be put to use to create wealth. It is a collective brainpower". Roos et al. (2005, p. 19) formulate the same concept in a more generic way: "Intellectual capital (IC) can be defined as all nonmonetary and nonphysical resources that are fully or partly controlled by the organization and that contribute to the organization's value creation".

Although there are many approaches to defining the ontology of organizational intellectual capital, the intersection of them yielded the canonical structure of *human capital, structural capital, and relational capital* (Andriessen 2004; Davenport 2005; Edvinsson and Malone 1997; MERITUM 2002; Ricceri 2008; Roos et al. 2005; Roos and Jutur 2005; Stewart 1997). *Human capital* represents knowledge, skills and competences of all employees. Human capital is fundamental for any company since it is the potential for knowledge creation and innovation. *Structural capital* reflects the functional and organizational structure of any company, including organizational culture. Structural capital can be defined, metaphorically, as being that part of organizational knowledge that remains within the company by the end of the day, when all employees left for home. *Relational capital* reflects the business relations the company has developed with its suppliers, customers and any other stakeholders. Although this structure is very simple and intuitive (which explains its popularity), it is confusing when it comes to measuring and reporting intellectual capital (see Chap. 8). Many metrics devised to measure intellectual capital contain indicators which have nothing to do with its intangible nature, like the number of computers, the number of people holding university diplomas, the number of research grants, or even the number of square meters of the production areas. Moreover, human capital, structural capital and relational capital are entities which are not independent entities, which makes the whole construct inconsistent with the logic of structuring.

There are also new approaches to finding basic and independent components of intellectual capital, again on the basis of the multi-field theory of organizational knowledge that we recalled in the previous chapters (Bratianu 2011, 2013, 2015; Bratianu and Orzea 2013). The *entropic* intellectual capital model contains, as basic entities: cognitive intellectual capital, emotional intellectual capital, and spiritual intellectual capital. In this perspective, human capital, structural capital, and relational capital appear only as a meta-construct of these basic entities. The entropic intellectual capital model can explain the organizational dynamics of the intellectual capital from a strategic perspective. The driving force of strategy is the shared vision of the company which, in turn, is a result of the visionary leadership, the main integrator of intellectual capital. The role of nonlinear integrators (Bratianu 2013, 2015) is to transform efficiently the potential intellectual capital into the operational intellectual capital. The model also contains the contribution of organizational learning and innovation in the renewal of intellectual capital. Thus, strategic knowledge as a resource is transformed by nonlinear integrators from its potential state into its operational state, and embedded in products and services put on the market. Also, strategic knowledge is transformed from its potential state into its co-operational form of dynamic capabilities, aiming at achieving competitive advantage. That means switching from the metaphorical interpretation of knowledge as stocks-and-flows to the multi-field interpretation based on the energy metaphor (Bratianu 2015). It is not the flow of knowledge through the company which opens the strategic perspective, but the entropic transformation of one form of knowledge into another which is able to shape the roadmap to competitive advantage and business sustainability. Clearly, this also leads to new difficulties for designing adequate metrics for intellectual capital, but, for sure, these metrics will appear essential in the future, due to their vital importance in measuring the success of any knowledge strategy.

3.2 Strategy as a Dynamic Notion

3.2.1 The Military Origin of the Concept

Since the creation of economic value is a (if not the most) important requirement for achieving success, this soon brings us to the idea of a strategy, because, as Grant (2016, p. 4) acutely says, "*strategy is about achieving success*". Consequently, when talking about the contribution of knowledge to value creation, it is necessary to talk about the connection between knowledge and strategy. We have already shown how knowledge, being considered an essential resource in the modern economy, acquires strategic importance in the literature. But how are these two notions connected in practical terms? Can knowledge be an ingredient of strategy formulation? Can it be the object of a strategy? In short: if knowledge is a *strategic weapon*, how kind of weapon is it?

When we talk about strategy, it is likely that any of us has an instinctive idea of what a strategy can be: we say that we have (or should have) a strategy in games like chess, or in agonistic sports, but also in our ordinary life (e.g. a strategy to find a job, to study, even to date a man or woman). But we need a clear definition of this notion, and here is where we've got a problem: although this term is one of the most frequently used in business, its definition is quite unstable.

The origins of the word are, undoubtedly, military: so, strategy is, first of all, about how to achieve success in a war. Etymologically, strategy comes from the Greek word "strategos" (στρατηγόσ) which means "who that leads an army". This is a first important point to notice: in the original idea, a strategy implies a *leadership*. We will later see how this fits the current notion that we have in the business environment. It is remarkable that while the military specialized literature has not used it as direct object of consideration until relatively recent times, its importance was very clear from the earliest times. The famous *Art of War* is considered the most ancient military treatise (fifth century BC) and is generally attributed to the Chinese general Sun Tzu (for an English translation: see Tzu 2015a, b). Its author doesn't explicitly define or adopt the term strategy, but he clearly underlines the importance to: (a) have the clearest idea of the goals; (b) have good knowledge of enemies, of their points of strength and weakness, of the war conditions, of the own forces; (c) know and plan the use of fighting techniques, tactics, and field positions; and (d) know how to employ leadership effectively. So, although there is no clear definition of strategy, this ancient text raises an essential point: to achieve success in war, there is a lot that a person must attempt to know. In a recent reinterpretation of Sun Tzu's book for business managers that has been made by Krause (2002), the words of the ancient Chinese General are summarized as follows: "Conquerors estimate in their temple before the war begins. They consider everything. The defeated also estimate before the war, but they do not consider everything. Estimating completely creates victory. Estimating incompletely causes failures. When we look at it from this point of view, it is obvious who will win the war" (Krause 2002, p. 10). So, it may be said that the perfect

Commander is also a perfect Strategist that is capable to analyze and predict any possible situations that can occur in the battleground.

A different view of military strategy is proposed in the famous work by Miyamoto Musashi, who is considered a legendary martial artist of seventeenth century Japan. In his *Book of five rings*, Musashi emphasized the essential role of learning and strategic thinking: "To learn technique is essential. However, strategic thinking has its own principles and they can be applied to anything that has to do with war and combat" (see Kaufman and Kaufman 1994, p. 14). For Mushashi, the power of a warrior should come from inside, from his mind, his heart and his spirit. Strategy means to win, but first the warrior must win in his mind before starting the battle. Technique is only supportive since strategy is thinking about the enemy, his strengths and vulnerabilities, and the context of the battle. Strategy means to belief in yourself and to wait for nor help from outside you. That means to understand the deep meaning of no-thing-ness: "you can come close to understanding no-thing by realizing that there is nothing outside of yourself that can ever enable you to get better, stronger, richer, quicker, or smarter. Everything exists. Seek nothing outside of yourself" (Kaufman and Kaufman 1994, p. 105). In short, the author underlines that strategizing also means to exploit oneself' emotional and spiritual strengths.

Probably, one of the first reasoned definitions of strategy was provided, later in history, by another popular expert, Carl von Clausewitz. In his "On War", written in 1848, he provides a concise definition: strategy is "the employment of the battle to gain the object of the war" (for an English translation, see: von Clausewitz 1873). So, there is no strategy without a goal. Earlier, in "*Principles of War*", written in 1812, he says that "This term means the combination of individual engagements to attain the goal of the campaign or war. If we know how to fight and how to win, little more knowledge is needed. For it is easy to combine fortunate results. It is merely a matter of experienced judgment and does not depend on special knowledge, as does the direction of battle."(for an English translation, see: von Clausewitz 1942). More precisely, strategy can't be simply an abstract exercise, but must be connected with reality and a capability to fit sudden changes: "Theory will therefore attend on strategy in the determination of its plans, or, as we may more properly say, it will throw a light on things in themselves, and in their relations to each other, and bring out prominently the little that there is of principle or rule.", and: "If we recall to mind . . . how many things of the highest importance war touches upon, we may conceive that a consideration of all requires a rare grasp of mind. A prince or general who knows exactly how to organize his war according to his object and means, who does neither too little nor too much, gives by that the greatest proof of his genius. But the effects of this talent are exhibited not so much by the invention of new modes of action, which might strike the eye immediately, as in the successful final result of the whole. It is the exact fulfillment of silent suppositions, it is the noiseless harmony of the whole action which we should admire, and which only makes itself known in the total result. The inquirer who, tracing back from the final result, does not perceive the signs of that harmony is one who is apt to seek for genius where it is not, and where it cannot be found." (von Clausewitz 1942). Von Clausewitz' idea of military strategy anticipates many issues that will be later considered: in particular, that strategy is a complex activity that implies a consideration of many different elements and issues, and that a "rare grasp of mind" is required to acquire enough knowledge of all them; in addition, it is the results that are eventually achieved that will decide that a strategy will have been really successful.

3.2.2 The Extension of Strategic Thinking to Business

Compared to the military context, the notion of strategy appeared in business quite late. In economics, the first explicit use of the term was probably introduced in Game Theory: depending on the type of game, a strategy is simply defined as "one of the given possible actions of a player" or "a complete plan of choices" (Turocy and von Stengel 2001, p. 3). It may be said that it is a restricted meaning, although it puts an emphasis on an important point: that an economic agent can have different actions or options to evaluate and to choose among.

It is just in the 1940s and 1950s, with the growth of big corporations, that management studies started to emphasize the importance of strategies. An often cited definition is that of Chandler (1962): "Strategy is the determination of the basic long-term goals of an enterprise, and the adoption of courses of action and the allocation of resources necessary for carrying out these goals." The fundamental keywords of this notion are: long-term *goals*, *adoption* (i.e. decision) of *courses of action*, and *resources*. It is a remarkably clear notion but, as Mintzberg et al. (1998) observe, it soon became evident that, for a company, formulating a strategy has much broader implications, which means that different views on strategy can be found:

- Strategy as plan, i.e. a course of action to achieve an *intended* set of goals.
- Strategy as pattern—a consistent pattern of past behavior, with a strategy *realized* over time rather than planned or *intended.*
- Strategy as position—locating brands, products, or companies within the market.
- Strategy as ploy—an action to outwit competitors.
- Strategy as perspective—i.e. it can be based on a "theory of the business" or it can be the extension of the mindset or ideological perspective of the organization/entrepreneur/strategist.

Here, the purpose is not to provide all the different perspectives or definitions that can be found, but the distinct ways these were and are used in the practice. The idea of strategy has been quite unstable over time, and this may be caused by the intrinsic *complexity* of the concept itself and the implications of its use: is it really possible to decide a goal, to evaluate different options and actions in advance, to select appropriate resources? Can strategists know enough of the company, environment, competitors, markets, etc. to do that appropriately?

Whatever definition of strategy one adopts, we may say that there are some general *assumptions* about the nature and characteristics of business and companies that need to be considered to talk about strategy (Bratianu and Bolisani 2015). A first element to consider is *heterogeneity*. If companies and markets were identical and operated in equal conditions, a strategy would have no meaning, because every company would instantly adopt the same course of action with no need to decide a distinct goal or direction to take (Davenport and Harris 2007; Johnson et al. 2011; Porter 1980, 1985). In short, the idea of a strategy implies an attempt to exploit the peculiar characteristics and unique or rare resources a company may have, compared to competitors, in the perspective of achieving a competitive advantage.

A second element is the orientation towards the *future*. A strategy implies a goal or an expected or desired *vision* of what the company should achieve or should be in the future. So, strategic thinking becomes a mental process oriented toward the future, which is not only based on factual data but also expresses our sense of life, work, competition, and winning. However, *future* is not a well-defined time domain and its perception is different in different cultures. Some people may consider that there is single and certain future which can be imagined as an extrapolation of the present time. For them, strategic thinking becomes an extension of the operational planning. Some people consider that there is a single future but it is populated with uncertain events and phenomena such that decision making will involve some risks. In that situation developing a strategic thinking is necessary. Also, there are managers for whom there could be multiple futures as a result of the high level of uncertainty and random interactions between events and phenomena. At the limit, future can be conceived as a chaotic environment where strategists look for some order or directions to follow toward some strange attractors. Finally, there can be executives that consider the future an opportunity that can be actively built, so that strategy becomes a positive intention and the manifestation of their strong willingness. These aspects will also be discussed in the next chapter.

A third element is that formulating a strategy involves and challenges all the *cognitive capabilities* of strategists, whoever they are in the company. "Strategizing [is] the judgment or imaginative response to what is NOT known, to the surprising, unexpected, incomplete, or illogical nature of what arises through our practice" (Spender 2014, p. 21). A strategy means projecting a company into a plausible future, which requires the capability to "see beyond".

Finally, under a perspective that considers *micro-foundations*, a strategy can be seen not as an outcome of a single mastermind from the top of the managerial hierarchy, but as an *integrated contribution of many people* working at different managerial levels and interacting in a dynamic social structure. The dynamics of *human and social interaction* become essential in determining the aggregated outcomes as dynamic capabilities of the firm (Teece 2009).

3.3 The Knowledge-Based View of Strategy

3.3.1 Classical Streams in Economics

In this section, we will discuss the evolution of the notion of strategy in the managerial literature. Clearly, the purpose is not that of providing a complete overview of the pertinent literature: this would be quite difficult and, in any case, well beyond the aims of this book. Rather, we intend to show how the concepts of knowledge and information are, in one form or another, central in the long lasting debate about strategy and strategic thinking. We will also underline how the evolution of the notion has also been affected by the evolving place of knowledge in the management literature.

To understand the evolution of the notion of strategy, it is useful to step back and verify how this concept is used in the "traditional" economic schools. With this unorthodox term, we want just to include some of the most important schools of thought in economics. Going back of more or less one century, we must first mention *neoclassical economics*. This term became popular in the 1950s (Aspromourgos 1986) but refers to earlier scholars, including Alfred Marshall and the marginalists, Carl Menger and the so-called Austrian School, and many others. Although the models of neoclassical economics were also criticized and contrasted, their influence in the economic thought was and is still central. Particularly important are the central assumptions about the behavior of economic agents (i.e. firms and consumers), that can be summarized as follows (Weintraub 2007): (a) agents have rational preferences regarding the possible choices, and these preferences can be clearly identified and valued; (b) firms' goal is maximization of profits, and consumers' goal is maximization of individual utility; and (c) agents act individually, but on the basis of the same full and relevant information. This has important implications: in the same conditions, firms would act the same way. So, a situation of perfect information leads to no strategy, as is well underlined by Knudsen (2011, p. 179): "Economics and strategic management have, for many years, been developing relatively independently of one another. [....] By basing their theorizing on the assumption that firms within the same industry are subjected to identical cost and demand conditions, economists have actually assumed a world in which heterogeneity among firms cannot occur. However, this heterogeneity is the very precondition for the domain of strategic management".

Along the same stream, but with important differences, we can place the studies of Industrial Organization. In particular, scholars add real-world complications to the model of perfect competition developed in neoclassical economics. A model that is used to explain firm performance and behavior is the so-called structure–conduct–performance paradigm, developed by Edward Chamberlin, Joan Robinson and Joe Bain. To put it in a nutshell, according to this paradigm, the structure of a market has a direct influence on the firm's economic conduct, which in turn affects its market performance. In other words, strategies of companies (i.e. the way they decide and plan their conduct) depend on the industry where they operate, which defines the structure of the market. So, again, it may be said that, at least in the same sector, firms may tend to have the same conduct (i.e. strategy) because they are exposed to the same information regarding the market conditions where they operate; while in different industries, companies may act differently.

In short, whenever information is perfect and symmetrically distributed among economic agents, there is no possibility of a strategy in strict sense or, better, there is no need of a strategy, i.e. an effort to define goals, plans and resources. Companies and entrepreneurs have simply to process the available information (about markets, prices, technologies, etc.) and to act consequently.

The main contribution of the Austrian economist Joseph Schumpeter is his explanation of economic systems as intrinsically dynamic. Differently from the traditional vision based on neoclassical influences, Shumpterer emphasizes the role of entrepreneurs, who compete not only by reacting to changes in price factors and demand (in a condition of symmetric and perfect information), but also by anticipating the behavior of competitors. For this, innovation is the main weapon: "Entrepreneurs are the subjects who can see new combinations of resources (rather than just adapting or improving existing combinations), and introduce such discontinuities so to create waves of innovation" (Buzzavo 2012). Creative destruction is the term often used to synthesize Schumpeter's perspective: a capability to proactively change the environment where the company operates. This has two important implications. First, Schumpeter's vision implies a recognition of *heterogeneity*: firms are different because entrepreneurs have different pieces of information, and different capabilities to process these and to create new knowledge that will enable their companies to succeed, by introducing and exploiting innovations, before or better than competitors. Second, there is an implicit notion of strategy that emerges here, i.e. a capability to "see beyond", to process "weak signals", and to decide goals that can be different from others'. A notion of strategy that, by using Cherp et al. (2007) words, is "informal and visionary", "largely intuitive", and mostly based on a unique leadership capability of the individual entrepreneur.

3.3.2 The Age of Strategic Planning

Strategic planning originated in the 1950s and 1960s of the past century, thanks to influential contributors such as Alfred Chandler, Igor Ansoff, and some others. As we have seen, before them, terms such as strategy and strategic planning were substantially absent in the economic and business management literature (Kiechel 2010). In those decades, the economic environment was subjected to important transformations. Big industries (among others, in particular: Chemical sector, Automotive, Consumer electronics, Household appliances, etc.) started to grow fast due to the impressive impulse of product innovations and to market demand that rapidly increased after the disasters of the II World War. Companies begun to grow in size and market coverage, and particularly the American big corporations. "Senior executives experienced increasing difficulty in coordinating decisions and maintaining control in companies that were growing in size and complexity" (Grant 2016). Terms such as "corporate planning", "long-term planning", "strategic planning" and similar, rapidly became popular, and a sort of "new science" of rational management made its appearance in both the professional context and in universities and business schools. New techniques of forecasting, cash flow analysis, rational decision making and project portfolio analysis were developed, and their use was based on the availability of reliable macro and microeconomic forecasts on markets, prices, market shares, costs and margins, etc.). The majority of US and European large companies set up their corporate planning departments, with the purpose to produce formal documents that defined long-term goals and plans of the company, expressed in key economic terms (i.e.: production, costs, revenues, market share, margins, etc.). This is the glorious age of *rational planning*, substantially based on a sort of pre-defined procedure that can be here described as follows: (a) Collection of information and pertinent data on company and markets; (b) definition of goals; (c) identification of alternative plans/policies; (d) evaluation of alternative plans/policies; (e) implementation of plans/policies; (f) Monitoring of effects of plans/policies. The main assumption is that companies are able to: (a) collect pertinent information, appropriate and sufficient for planning; (b) analyze this information rationally, especially in terms of its implications for future actions and goals, and (c) take consistent decisions about plans and courses of actions. Again, the basic point is that, like the classic economic theories, it is assumed that decision makers can really acquire complete information about the relevant economic elements, although it is now recognized that these elements can be different from a company to another (i.e. different markets, products, production systems, etc.) and, in addition, what also counts is the capability of strategic departments to analyze this information with appropriate tools and to derive useful knowledge for executives and for their decision making.

The following decades experienced a crisis of rational planning in corporations and a new era of economic instability. A period of economic stagnation started, due to various reasons including a change in market needs, high inflation and increasing unemployment. Also, international competition of newcomers (especially Korean and Japanese corporations) challenged the certainties of Western-based strategic approaches. "The new turbulence meant that firms could no longer plan their investments and resource requirements 3–5 years ahead—they couldn't forecast that far ahead" (Grant 2016, p. 13). We can reformulate this case saying that companies had a new awareness: it was virtually impossible to collect sufficient information for rational planning, and companies learnt that they simply couldn't "know enough" for that. Since the dynamics and turbulences of markets and technologies challenged the possibility to plan strategy rationally, some studies proposed to consider the effects of the overall uncertainty on managerial decisions: uncertainty implies that companies never have all the information that would be required to rationally set plans in advance; dynamic changes imply that new scenarios may emerge over time. Indeed, some time earlier, a prophetic sign of the fall of faith in rational planning had already appeared in the works of Nobel Prize Herbert Simon (Simon 1949): Simon contrasted the classic assumption that economic agents (and human beings in general) can be approximated as rational agents. His notion of *bounded rationality* accounted for the fact that perfectly rational decisions are simply not feasible, due to the intrinsically insufficiency of information for a vast majority of practical decision problems, and the finite processing capabilities of individuals or even automated systems. All this hinders the possibility to acquire enough knowledge for rational decision making.

So, an important notion was proposed, that of *emergent strategy* (Mintzberg and Waters 1985), which contrasts that of rational planning. Rational planning implies an explicit and rational formulation of goals, plans and means in advance, as originated from precise intentions of the company. The strategy is decided by central leadership, is progressively articulated in detailed tasks for the different parts of a company, and is implemented by means of formal controls in a top-down logic. As mentioned, the essential pre-conditions for this approach to work properly are that the internal and external environment can be "known", i.e. it is controllable or predictable; also, the entire company has full understanding, adhesion and acceptance of the specific tasks and processes that are required by strategic goals and plans.

The new view of Mintzberg and colleagues attempts to explain how "things really work" in a world that is intrinsically affected by uncertainty. Emergent approach to strategy means the recognition that actions result to be consistent over time, but in the absence of ex-ante intentions, clear leadership, or predefinition of goals or plans. In other words, in an emergent approach, goals and plans of a company just result from an ex-post formalization and co-ordination of actions, decisions and tasks that have proven to be effective and beneficial to the organization. The idea of emergent strategy doesn't imply absence of planning efforts *per se*, but it assumes that, since too little information can be available in advance, strategists can only, in the best case, learn by experience, in a trial-and-error process. Of course they can try to formulate a strategy, but their supposed rational approaches are not based on a consistent base of knowledge: it is only those courses of action that appear to provide better results that can be selected and, maybe, later declared as "the official company strategy". Although, as Mintzberg and Waters (1985) recognize, a purely emergent approach may be an abstract "ideal-type", it is not difficult to find situations, in the real life of companies, that are more or less close to that theoretical definition.

3.3.3 Planning Strikes Back: Competitive Strategies and Resource-Based Views

Indeed, the notion of emergent strategy leaves little room to any planning effort, and may somewhat lead to a "negation" of the significance itself of strategy as a notion, as was substantially proposed by Weick (1987, p. 221–222): "If pressed to define strategy, I am tempted to adopt DeBono's [...] statement that 'strategy is good luck rationalized in hindsight' but I am also comfortable with a definition much like Robert Bulgermanns' [...] namely ' strategy is a theory about the reasons for past and current success of the firm".

Regardless the appeal of Weick's fascinating provocation, it is also clear that companies can't just move randomly in the dark, hoping that something good happens soon or later. So, during the 1980s and 1990s, new approaches to strategy appeared in the literature. Michael Porter's notion of *competitive strategy* implies that a company seeks to achieve a superiority compared to its *competitors* in terms of price or product differentiation (Porter 1980, 1985). Porter's popular models, like the Value Chain, the 5 forces, or the Generic strategies, are substantially based on an important assumption: it is not essential that companies *know everything* about themselves and the environment so that they can plan their choices and actions rationally, but it is more important to *achieve knowledge* about the forces that affect the particular competitive environment where a company operates. It is a sort of reduction in planning complexity, although strategists are still asked to collect enough information about what should be known to set a competitive strategy. In a sort of contingency view, Porter assumes that what is really important is that a company learns how its strategy can fit the environmental contingent situation (Grant 2016).

A shift in strategic thinking is represented by the *resource-based view of the firm* that we have already mentioned in the earlier chapters. Here, the focus is on *internal resources* as the key element that can enable a company to make profit, rather that *external competition*. In other words, the basis is to understand the internal resources that can't be replicated or imitated easily by competitors, and to transform these into competitive advantage (Barney 1991). In terms of knowledge, there are important implications. First of all, planning a strategy needs an exercise of "self-understanding": in other words, strategists need to collect appropriate information and achieve knowledge of internal resources, either tangible and intangible, and about how to exploit them. Second, among intangible resources, knowledge makes its important appearance: as we have recalled, the so-called *knowledge-based theory of the firm* (Grant 1996) directly stems from the resource-based view, and sees companies as institutions "for integrating knowledge". Placing knowledge at the core of firms' activity implies that planning a strategy for a company implies to formulate a strategy regarding the production and processing of knowledge. Knowledge becomes not only an input of strategy formulation, but also its object.

The resource-based view of the firm can be framed into a "wider class of theories, variously described as 'capabilities', 'resource-based', or 'competence-based' theories of the firm" (Hodgson 1998). Among these, the *evolutionary view of the firm* has a special place. Developed after the seminal work by Nelson and Winter (1982), the theory focuses on the mutations in technology and routines that companies may need to implement for facing changes in the environment that may threat the 'status quo' and the capability to remain in a state of equilibrium. The main point is that change can occur constantly in the economy, so for companies a sort of continuous evolutionary process must be in action. Similar to a Darwinian process in biology, the theory analyzes the dynamics of economy by assuming an implicit mechanism of generation of variations and selection of the organizational solutions that "best fit the environment". For companies to survive, this means ability to modify their own routines and capabilities: in other words, a strategy to *learn*.

Senge's (1990) notion of *learning organization* moves further into this stream of thought. In a learning organization, given the awareness that change is essential for survival, members are continuously stimulated to learn and, by this way, to prepare change and innovation for finding better dynamic fit to the external environment. For a learning company, a strategy also includes *a strategy of change* by means of continuous learning. A concept that has become popular in business schools since the 1990s and that extends that of learning organization is *strategic agility*. This term, that recalls similarly popular words like Agile manufacturing or Agile software development, puts and emphasis on the capability of a firm to respond to environmental change by adapting its configuration, by maintaining and adapting goods and services to meet customer demands, adjusting to the changes in environment, taking advantage of human resources and their learning capability, etc. According to Roth (1996, p. 32), strategic agility "requires a metamorphosis from the organization as mechanistic "working machine" to [...] an organic, accelerated learning organization that produces shared knowledge as a key by-product". In short, a strategic attitude to agility means a shift from rational planning to a new form of strategic thinking that "requires a delicate balancing act between holding fast to your vision of winning while adjusting to the constant upheavals in the world around us" (Green 2011). All this "requires economies of knowledge through accelerated enterprise-wise learning" (Roth 1996, p. 30). In short, in all these different but converging contribution, the central place of knowledge is clear.

3.3.4 Strategy as Practice

A recent important contribution to strategic thinking is that of *strategy-as-practice*. More than a definition of strategy, it is a new research perspective—or research agenda—whose aim is "to break through the economics-based dominance over strategy research" (Jarzabkowski 2005, p. 3) and that combines different fields from sociology, organization theory, and systems thinking. "The focus of this approach is on strategy as a social 'practice', on how the practitioners of strategy really act and interact. From the perspective of strategy as practice, the key question is: what does it take to be an effective strategy practitioner?" (Whittington 1996, p. 731). Practice is concerned with the process of strategizing (Spender 2014), which is seen as an interactive and iterative process between strategic thinking, strategy elaboration, and strategy implementation. It is an accelerating convergence between the desirable future and the present, i.e., between the framework of what one rationally chooses and what comes from real-life experience. In companies, this means integrating these two views across all levels of the organizational structure, from individuals and micro activities to middle management and top management of the company. To some extent, the Strategy-as-practice approach is "part of a broader practice turn in contemporary social theory and the management sciences over the past 20 years. It has been imported into such diverse management fields as technology, knowledge management, organizational learning and accounting"; also, "it may be seen as the culmination of broader shifts in strategic management", because "it responds to challenges and issues raised in strategy process, resource-based view and dynamic capabilities research" (Jarzabkowski 2005, p. 3).

The key concept of this approach is not the noun "strategy", but the verb "strategizing", which is context dependent and sensitive to cultural values. This also proves to be stimulating for managers with creative minds. Essential elements of this new view are as follows (Jarzabkowski 2005): (a) strategy is *situated activity*, i.e. it is a process that shapes and is shaped by the environment where it occurs; (b) strategy is always *under construction*, (c) strategy is a *distributed activity*, i.e. it is not the results of separate reasoning made by special people in the company (i.e. the *strategists*) but the combination of collective contributions coming from all the parts of an organization, and (d) executives are not omniscient and all-powerful actors that arrange the whole activities of the company, but rather *mediating agents* that can combine and re-formulate all the meaningful suggestions that come from the organization. A general feature of this approach is that strategy becomes a sort of *social activity*, and strategizing means to reflect on issues e.g.: what do we already know in our company? Who knows what in our organization? What will we need to know in the future?

3.4 Lessons from a Long History

As can be observed from the analysis conducted in the previous section, it is possible to read the evolution of the concept of strategy on the basis of the notion of knowledge, of its basic ingredient (information), or of other concepts (e.g. learning) that are related to them. Table 3.1 proposes an outline of what discussed in Sect. 3.3.

This brief discussion makes it possible to make some key points that will be important in the rest of the book. First of all, knowledge and its related concepts are, in a way or another, essential to understand the definitions of strategy and the perspectives under which this concept has been seen in the economic and managerial literature. Second, the increasing awareness that companies are complex organizations that act in a turbulent environment, affected by uncertainty and unpredictability, strengthens the recognition that knowledge and related processes (i.e.: knowledge creation, sharing, learning, etc.) have a central place in the work of "strategists" (or of anybody that must take strategic decisions in a company). Third, that knowledge is not simply a more or less complex "aggregation of data", but is intertwined with mental processes, attitudes, emotions, social behaviors, and even values of people. This is why it is necessary to adopt a new viewpoint of strategic planning that includes a capability to formulate a "knowledge strategy" or a "knowledge management strategy", as will be discussed in the following chapters.

Table 3.1 Evolution of the concept of strategy and role of knowledge

Literature stream	Role of strategy	Role of knowledge, information, and related concepts
Traditional economics	Absent or marginal	Perfect knowledge/information possessed by rational agents reduces possibility of autonomous behaviors by companies
Schumpeter	Creative destruction: heterogeneous entrepreneurs can autonomously decide the direction taken by their companies	Entrepreneurs can have distinct knowledge and cognitive capabilities
Strategic rational planning	Strategy is the rational planning of future activities, especially for big corporations in stable markets	Strategists rationally acquire and elaborate knowledge of external environment and of possible strategic choices
Simon's bounded rationality	Complete rationality in strategic planning is virtually impossible	Strategists must be aware of their insufficient cognitive capability even in stable conditions
Emergent strategies	Rather than planned or *intended*, a strategy can be *realized* and understood only over time	Uncertainty and environmental turbulence implies awareness of impossibility to acquire and process enough knowledge for formulating stable plans in advance
Porter's competitive strategy and "5 forces"	Strategy focuses on value creation to compete with others in a specific environment	Importance of knowledge about how to generate value better than direct competitors
Resource-based view	Strategy is based on exploitation of a company's key resources	Importance of "knowledge of self" (what we can do, what we know we can do, etc.)
Evolutionary view	Strategy as capability to change in reaction to the environment	Capability to learn from experience
Learning organization and strategic agility	Learning as a systematic strategy not only to react but also to anticipate changes	Learning by doing, learning by experimenting
Strategy as a practice	Strategy is a social process	Strategy is based on integration of distributed knowledge

3.5 Conclusion

The old saying "Knowledge is power" can be shaped into the new formulation "Knowledge is a strategic resource" to enhance its potential contribution to achieving competitive advantage. Since, for many people, the concept of *resource* refers only to stocks available for being used within a given process, in the first part of this chapter we emphasized the role of knowledge both as a *resource* and as a *strategic capability*. That means to address both dimensions of knowing about any production process—*know what* and *know how*—in a *time* perspective. By similarity with Porter's value chain analysis, it is important to perform a knowledge value chain analysis and to decide how to increase the economic value of each sequence of knowledge processing. That is necessary especially for the knowledge intensive business services where the density of knowledge per unit of production is very high. The result of such an analysis would be a decision of the adequate ratio between the outsourcing and insourcing activities. Big data and cloud computing become very attractive for outsourcing many knowledge processing sequences. Strategic knowledge is closely related to the organizational intellectual capital. Knowledge represents the main ingredient of intellectual capital, and its strategic feature suggests a detailed analysis of the whole phenomenon. Although intellectual capital has been developed mainly on the economic side of business and knowledge management has been developed from the managerial perspective of an organization, both concepts can be interpreted as two sides of the same coin.

The second part of the chapter focuses on the *dynamic semantic* of the concept of *strategy* and its consequences on the managerial practice. It is interesting to note the military origin of the concept and how its combat or war attributes can be mapped onto the business domain. If the main goal in war is to destroy your enemy, in business the goal is achieving competitive advantage. Strategies have been developed to achieve that goal. In the first stages of economic competition, strategies were more or less extrapolations of managerial planning techniques based mostly on a deterministic logic. With the increasing competition and future uncertainties due to changeable business environment, purely rational strategic planning has lost its appeal, and new notions such as emergent strategy have become popular. Finally, strategy-as-practice appeared recently as a new perspective of looking at turbulent business environment and using efficiently the lessons learned by managers from their direct business experience.

It is important to emphasize the fact that knowledge has been always used in developing business strategies, but in the beginning this has been limited only to economic aspects and rational planning. The new turbulent environment demonstrated that decision making is subjected to bounded rationality, so accepting irrationality is part of the game.

References

Andriessen, D. (2004). *Making sense of intellectual capital: Designing a method for the valuation of intangibles*. Amsterdam: Elsevier.

Aspromourgos, T. (1986). On the origins of the term 'neoclassical'. *Cambridge Journal of Economics, 10*(3), 265–270.

Augier, M., & Teece, D. J. (2007). Perspectives on research and development: Organizing and managing innovation in a (global) knowledge-based economy. In K. Ichijo & I. Nonaka (Eds.), *Knowledge creation and management: New challenges for managers* (pp. 198–214). Oxford: Oxford University Press.

Barney, J. (1991). Firm resources and sustained competitive advantage. *Journal of management, 17*(1), 99–120.

Barney, J., & Hesterly, W. S. (2012). *Strategic management and competitive advantage: Concepts and cases*. New York: Pearson.

Bratianu, C. (2011). A new perspective of the intellectual capital dynamics in organizations. In B. Vallejo-Alonso, A. Rodriguez-Castellanos, & G. Arregui-Ayastuy (Eds.), *Identifying, measuring, and valuing knowledge-based intangible assets: New perspectives* (pp. 1–21). Hershey: IGI Global.

Bratianu, C. (2013). Nonlinear integrators of the organizational intellectual capital. In M. Fathi (Ed.), *Integration of practice-oriented knowledge technology: Trends and perspectives* (pp. 3–16). Heidelberg: Springer.

Bratianu, C. (2015). *Organizational knowledge dynamics: Managing knowledge creation, acquisition, sharing and transformation*. Hershey: IGI Global.

Bratianu, C., & Bolisani, E. (2015). Knowledge strategy: An integrated approach for managing uncertainty. In A. Garlatti & M. Massaro (Eds.), *Proceedings of the 16th European Conference on Knowledge Management, University of Udine, Italy, 3–4 September 2015* (pp. 169–177). Reading, MA: Academic Conferences and Publishing International limited.

Bratianu, C., & Orzea, I. (2013). The entropic intellectual capital model. *Knowledge Management Research and Practice, 11*(2), 133–141.

Brown, J. S., & Duguit, P. (1998). Organizing knowledge. *California Management Review, 40*(3), 90–110.

Buzzavo, L. (2012). Strategy innovation as business model reconfiguration. Working Paper Series, 2/2012. University of Venice, Ca' Foscari.

Chandler, A. (1962). *Strategy and structure: Chapters in the history of industrial enterprise*. New York: Doubleday.

Cherp, A., Watt, A., & Vinichenko, V. (2007). SEA and strategy formation theories: From three Ps to five Ps. *Environmental Impact Assessment Review, 27*(7), 624–644.

Christensen, C. M. (2003). *The innovator's dilemma*. New York: Harper Business Book.

Davenport, T. H. (2005). *Thinking for aliving: How to get better performance and results from knowledge workers*. Boston: Harvard Business School Press.

Davenport, T. H., & Harris, J. G. (2007). *Competing on analytics: The new science of winning*. Boston: Harvard Business School.

Downes, L., & Nunes, P. (2014). *Big bang disruption: Business survival in the age of constant innovation*. London: Penguin/Random House.

Drucker, P. F. (1969). *The age of discontinuity*. Piscataway, NJ: Transactions Publishing.

Drucker, P. F. (1993). *Post-industrial society*. New York: Harper Business.

Edvinsson, L., & Malone, M. (1997). *Intellectual capital: Realizing your company's true value by finding its hidden brainpower*. New York: HarperCollins.

Grant, R. M. (1996). Toward a knowledge-based theory of the firm. *Strategic Management Journal, 17*(S2), 109–122.

Grant, R. M. (1997). The knowledge-based view of the firm: Implications for management practice. *Long Range Planning, 30*(3), 450–454.

Grant, R. M. (2016). *Contemporary strategy analysis: Text and cases edition*. New York: Wiley.

Green, H. (2011, May 28). *Shifting from strategic planning to strategic agility*. Forbes.

Hodgson, G. M. (1998). Evolutionary and competence-based theories of the firm. *Journal of Economic Studies, 25*(1), 25–56.

Holsapple, C. W., & Singh, M. (2003). The knowledge chain model: Activities for competitiveness. In C. W. Holsapple (Ed.), *Handbook on knowledge management* (Vol. 2, pp. 215–252). Berlin: Springer.

Ichijo, K. (2007). The strategic management of knowledge. In K. Ichijo & I. Nonaka (Eds.), *Knowledge creation and management: New challenges for managers*. Oxford: Oxford University Press.

Jarzabkowski, P. (2005). *Strategy as practice: An activity based approach*. Thousand Oaks, CA: Sage.

Jenster, P., Pedersen, H. S., Plackett, P., & Hussey, D. (2005). *Outsourcing – insourcing: Can vendors make money from the new relationship opportunities?* New York: Wiley.

Johnson, G., Whittington, R., & Scholes, K. (2011). *Exploring strategy: Text & cases* (9th ed.). London: Prentice Hall.

Kabir, N., & Carayannis, E. (2013, October 21–23). *Big data, tacit knowledge and organizational competitiveness* (p. 220). Proceedings of the 10th International Conference on Intellectual Capital, Knowledge Management and Organisational Learning, George Washington University, Washington, DC.

Kaufman, S., & Kaufman, S. F. (1994). *The martial artist's book of five rings: The definitive interpretation of Miyamoto Musashi's classic book of strategy*. North Clarendon: Tuttle Publishing.

Kiechel, W. (2010). *The lords of strategy: The secret intellectual history of the new corporate world*. Boston: Harvard Business Press.

Knudsen, C. (2011). Theories of the firm, strategic management. In C. A. Montgomery (Ed.), *Resource-based and evolutionary theories of the firm: Towards a synthesis*. New York: Kluwer.

Krause, D. (2002). *Sun Tzu: The art of war for executives*. London: Nicholas Brealey Publishing.

Lamont, J. (2012). Big data has big implications for knowledge management. *KM World, 21*(4), 8–11.

MERITUM. (2002). *Guidelines for managing and reporting of intangibles*. Madrid: Fundacion Airtel Movil.

Minelli, M., Chambers, M., & Dhiraj, A. (2012). *Big data, big analytics: Emerging business intelligence and analytic trends for today's businesses*. New York: Wiley.

Mintzberg, H., & Waters, J. A. (1985). Of strategies, deliberate and emergent. *Strategic Management Journal, 6*(3), 257–272.

Mintzberg, H., Ahlstrand, B., & Lampel, J. (1998). *Strategy safari: A guided tour through the wilds of strategic management*. New York: The Free Press.

Nelson, R. R., & Winter, S. G. (1982). *An evolutionary theory of economic change*. Cambridge: Harvard University Press.

Nonaka, I. (1994). A dynamic theory of organizational creation. *Organization Science, 5*(1), 14–37.

Nonaka, I., & Takeuchi, H. (1995). *The knowledge-creating company: How Japanese companies create the dynamics of innovation*. Oxford: Oxford University Press.

North, K., & Gueldenberg, S. (2011). *Effective knowledge work: Answers to the management challenge of the 21st century*. Bingley: Emerald.

Pasher, E., & Ronen, T. (2011). *The complete guide to knowledge management: A strategic plan to leverage your company's intellectual capital*. New York: Wiley.

Porter, M. (1980). *Competitive strategy*. New York: Free Press.

Porter, M. (1985). *Competitive advantage: Creating and sustaining superior performance*. New York: Free Press.

Prusak, L. (2009). Introduction to knowledge in organizations. In L. Prusak (Ed.), *Knowledge in organizations*. London: Routledge.

Ricceri, F. (2008). *Intellectual capital and knowledge management: Strategic management of knowledge resources*. London: Routledge.

Roos, G., & Jutur, S. (Eds.). (2005). *Intellectual capital management: A strategic perspective*. Hyderabad: The ICFAI University Press.

Roos, G., Pike, S., & Fernström, L. (2005). *Managing intellectual capital in practice*. Amsterdam: Elsevier.

Roth, A. V. (1996). Achieving strategic agility through economies of knowledge. *Planning Review, 24*(2), 30–36.

Rother, M. (2010). *Toyota kata: Managing people for improvement, adaptiveness, and superior results*. New York: McGraw-Hill.

Senge, P. (1990). *The fifth discipline: The art and science of the learning organization*. New York: Currency/Doubleday.
Simon, H. A. (1949). *Administrative behavior. A study of decision-making processes in administrative organisation* (Fourth Printing). London: Macmillan Company.
Spender, J. C. (1996). Making knowledge the basis of a dynamic theory of the firm. *Strategic Management Journal, 17*(S2), 45–62.
Spender, J. C. (2014). *Business strategy: Managing uncertainty, opportunity, & enterprise*. Oxford: Oxford University Press.
Stewart, T. (1997). *Intellectual capital*. New York: Doubleday.
Sveiby, K. E. (2001). A knowledge-based theory of the firm to guide in strategy formulation. *Journal of Intellectual Capital, 2*(4), 344–358.
Taleb, N. N. (2008). *The black swan: The impact of the highly improbable*. New York: Penguin Books.
Teece, D. J. (2009). *Dynamic capabilities & strategic management*. Oxford: Oxford University Press.
Triplett, J. (1999). The Solow productivity paradox: What do computers do to productivity? *Canadian Journal of Economics, 32*(2), 309–334.
Turocy, T. L., & von Stengel, B. (2001). Game theory: CDAM Research Report LSE-CDAM-2001-09. Texas A&M University & London School of Economics, Tech. Rep.
Tzu, S. (2015a). The art of war (public domain). Online. Retrieved January 27, 2017, from http://classics.mit.edu/Tzu/artwar.html
Tzu, S. (2015b). *The art of war*. New York: Sheba Blake Publishing.
von Clausewitz, C. (1873). *On war* (J. J. Graham, Trans.). London: N Trübner.
von Clausewitz, C. (1942). In H. W. Gatzke (Ed. & Trans.), *Principles of war*. London: The Military Service Publishing Company.
Weick, K. E. (1987). Substitutes for strategy. In D. Teece (Ed.), *The competitive challenge* (pp. 211–233). New York: Harper & Row.
Weintraub, E. R. (2007). Neoclassical economics. In D. R. Henderson (Ed.), *The concise encyclopedia of economics* (online). Retrieved from http://www.econlib.org/library/Enc1/NeoclassicalEconomics.html
Whittington, R. (1996). Strategy as practice. *Long Range Planning, 29*(5), 731–735.

Understanding the Future for Strategy Formulation

4

The purpose of this chapter is to explain our perception of *time* by using metaphorical thinking and to show how we understand the concept of *future* within the framework of time, complexity and uncertainty. Strategies are built for future actions and understanding the nature and the content of future becomes important. Human mind developed, during its historical existence, a series of metaphors able to suggest new semantic dimensions of time and its role in structuring the future. Among all these metaphors those based on *space* are essential since time and space have been integrated by scientists in a complex n-dimensional space concept. Time specialization became, thus, the main cognitive pattern in dealing with time and the future. Due to the complexity of future it is important to explain the way we correlate variables describing events and phenomena, which means to address the linearity and nonlinearity paradigms. The chapter ends up by presenting the semantic dynamics of uncertainty and its role in defining probable futures, where we define corporate strategic objectives and design strategies able to achieve them.

4.1 Time Perception

4.1.1 Inertial Thinking

We learn from physics that *inertia* is a phenomenon which characterizes the tendency of any physical object to keep its existing state of motionless or that of a uniform motion in straight lines, until an external force changes it. That means that inertia represents a form of resistance to any possible change or transformation of the given state. The inertial or resisting force is proportional to the mass of that physical body and opposes to the external force acting upon it. The body changes its state only if the external force is larger than the inertial force. The meaning of the inertial force has been extended beyond the physical world and generalized to any tendency of doing nothing or remaining unchanged. Thus, there are many synonyms sharing that meaning like inactivity, inaction, apathy or lethargy.

E. Bolisani, C. Bratianu, *Emergent Knowledge Strategies*, Knowledge Management and Organizational Learning 4, DOI 10.1007/978-3-319-60657-6_4

The essence of the inertia phenomenon is that it is timeless, which means that time is not a part of the model governing the state of an inertial object. We can extend this idea to *thinking models* (Fauconnier and Turner 2002; Frith 2007; Senge 1999) which represent ways of understanding and explaining the world we are living in. That model of thinking, which contains no time reference or correlates timeless variables, is called *inertial thinking* (Bratianu 2007, 2015a). When we return back home after a heavy working day we do expect everything to be in the state we left it in the morning. No change and no transformation. The walls of the house are there, the furniture is there and all the little things spread out through the house have not changed their state. That generates a feeling of stability and security. We need that feeling and that is why we value inertial thinking. When we think about physical objects, this is almost natural since they do not have the power of moving. However, when we think about the social environment and our business, the whole picture changes due to so many forces acting upon us and many events happening around us. Inertial thinking does not contain time in its structure and yields a timeless existence feeling. That means that it is not able to accept change, a motion or transformation done in time and with time. It will oppose any change and will become a resistance to change. Inertial thinking is a good friend of stability but a great enemy of change. When inertial thinking is dominant in our decision making, we would have the natural tendency to oppose any change.

In organizations, change is necessary for adapting continuously to the changing external business environment and for implementing strategies. However, *inertial thinking* opposes to change. For instance, when a successful business formula yields good economic results, managers will try to apply it again without any change even if the external business environment and the initial conditions are subjected to some changes. As Rumelt (2012, p. 203) remarks, "An organization's greatest challenge may not be external threats or opportunities, but instead the effects of entropy and inertia". Management has been invented to reduce the organizational entropy by designing structures, procedures and routines in order to increase efficiency and maximize the profit. Once they are created and put into operations, they have the tendency to preserve as they are and to accept only small improvements without changing their nature and functions. They encapsulate knowledge and intelligence reflecting the "way things are done" and the dominant managerial logic: "These routines not only limit action to the familiar, they also filter and shape manager's perceptions of issues. An organization's standard routines and methods act to preserve old ways of categorizing and processing information" (Rumelt 2012, pp. 203–204).

In 1978 in USA, the deregulation of airline transport created almost a shock for all air travel companies (Rumelt 2012). All the routines for managing such companies and understanding competition had been based on strong regulation. Changing the legislation created a new context for competition and successful business. However, many companies managed their business for some years according to the old routines remaining captive to inertial thinking. "From 1979 to 1983, the majors kept enacting the old rules. In 1981, United, American, and Eastern together lost $240 million, while all of the shorter-haul carriers (Delta, Frontier, USAir, and so on) made a profit. Over the next two decades, only Southwest would be consistently profitable" (Rumelt 2012, p. 207). More than routines, organizational culture can be a huge resistance to change since it integrates norms, believes, traditions and values. It embodies spiritual knowledge as shared values which reflect the essence of any corporate managerial philosophy. Changing organizational culture is rather difficult due to the fact that it contains intangible entities which cannot be seen, touched or address directly. Kotter (1996, 2008) and Kotter and Cohen (2002) demonstrate that inertial thinking embedded in organizational culture represents a great resistance to change, reason for which any change process should not start with changing culture, as some old models claimed it. According to Kotter (1996, p. 156), "Culture is not something that you manipulate easily. Attempts to grab it and twist it into a new shape never work because you can't grab it. Culture changes only after you have successfully altered people's actions, after the new behavior produces some group benefit for a period of time, and after people see the connection between the new actions and the performance improvement". Another factor contributing to inertial thinking is the feeling of complacency, which comes mostly from the unconscious emotion (Kotter 2008). In the dictionary, we find that complacency is a feeling of satisfaction with yourself or with a given situation, so that you do not think any change is necessary. That means that inertial thinking integrates rational knowledge, emotional knowledge and spiritual knowledge, contributing essentially to maintaining a given behavior of people or a given state of an organization. The perception of the future in such a situation is just an extension of the present time without any thinkable change. There is no uncertainty about that future and there is no need for designing any strategy since everything is known and under control. Strategies have been created to deal with uncertainty, knowledge incompleteness and all the possible risks generated by them. Inertial thinking makes all of these strategies unnecessary.

4.1.2 Dynamic Thinking

The *dynamic thinking* model incorporates time as a fundamental variable (Bratianu 2007). That means that dynamic thinking accepts change as a natural phenomenon. However, dynamic thinking reflects the Newtonian logic which is based on reversible processes. A process is called *reversible* when it comes back to the initial state going through the same equilibrium states. If a process does not have that property, then it is an *irreversible* process. A reversible process is an ideal process conceived as a simplification of the natural phenomena in order to yield a practical solution to the mathematical model. For illustration we may consider the well-known formula for the average speed of a car driven from city A to city B. If the distance between the two cities is S and the time necessary for the car to cover that distance, then the average velocity is given by the Newtonian formula: $V = S/T$. That formula is valid also for the car to come back to city A considering that there are the same equilibrium states, which happens only in a theoretical approximation. In real life there are no reversible processes. All processes are irreversible.

It is also interesting to note that time T in the above formula has only a quantitative dimension expresses by units of measurement of time (i.e. seconds, minutes, hours, etc.). It has no qualitative dimension which means that it contains no ordered structure of past, present and future. Even though we may consider that there are these three different qualitative zones of the time variable, they do not have a well-defined order or sequence of developing. Consider the following conceptual experiment. You live in London and leave your house early in the morning to drive toward Oxford, where you spend the rest of the day. Before you start driving the car you are in London in your present time, and Oxford represents the future. When you reach Oxford you are, again, in your present time and London represents your past. If you want to returned back to London and consider the whole process to be reversible, then London becomes your future and Oxford transforms into your past. Thus, past, present and future interchange their positions like in a circular motion. That is due to the basic hypothesis of a reversible process, which eliminates all resistances to change and transforms a real process into an ideal one. Since you start from London we may consider it the point zero of your travel. When you come back to London in a reversible process, you reach the initial state which means point zero. That is moving like in a circle. Thinking in circles is characteristic to Chinese people who have the quest for harmony. When Nisbett (2003, p. 14) analyzes the role played by Taoism in the Chinese way of thinking, he remarks: "Returning—moving in endless cycles—is the basic pattern of movement of the Tao".

Dynamic thinking is capable of understanding change but only within reversible processes. That means that any system may undergo change in the direct and reverse order such that the system recovers its initial state of equilibrium with the given context. In such a situation, any possible future may disappear into a past when the system returns to its original state. Time flows in circles and changes follow anticipated cycles. There is no uncertainty putting pressure on decision making and there is no need for strategies. In such a hypothetical world, the pattern of decision making is based on the Newtonian logic and laws of physics.

4.1.3 Entropic Thinking

In real life, all processes are irreversible. They evolve without being able of coming back to their initial states. For instance, in the mental experiment we discussed about somebody leaving London in the morning to drive to Oxford where to spend the rest of the day, coming back to London occurs in the evening which means a different time sequence with respect to morning. In real life, it is impossible to return back home going through exactly the same states of equilibrium like reversing completely all the time sequence. Real processes in nature and in society are irreversible and *time* becomes a fundamental variable reflecting a continuum structured in three distinct zones—past, present, and future—which are oriented from the past toward the future.

A key concept in describing irreversible processes is *entropy* which has been introduced in science by Rudolf Clausius in relation to the second law of thermodynamics. According to Atkins (2010, p. 37), "The second law is of central importance in the whole of science, and hence in our rational understanding of the universe, because it provides a foundation for understanding why any change occurs. Thus, not only is it a basis for understanding why engines run and chemical reactions occur, but it is also a foundation for understanding those most exquisite consequences of chemical reactions, the acts of literary, artistic, and musical creativity that enhance our culture". Although the concept of entropy has been introduced in thermodynamics to measure the quality of energy, its semantic has been extended to reversible and irreversible processes, to order and disorder and to many other phenomena which can be described by a distribution of probabilities (Ben-Naim 2012; Georgescu-Roegen 1999; Handscombe and Patterson 2004). In real processes, time has a direction from *past* toward *present*, and from *present* toward *future*. Future and past are not interchangeable like in the dynamic thinking; they represent distinct sequences in the lifecycle of any process. We may show this time orientation as follows:

Past>>>Present>>>Future

This orientation comes from irreversibility of processes, from the real impossibility of coming back to the starting point at any time. Scientists say that entropy introduces a *time arrow*. Understanding irreversible processes and thinking along the time direction toward the future means to develop the capability of *entropic thinking*.

Future does not exist in the physical world. It exists only in our minds as a potential complex of events and phenomena. We can construct that image of the future if and only if we have developed entropic thinking which allows us to direct our thinking toward the future along our irreversible timeline. People who are able to have such a vision about the future are called *visionary* people. Thomas Alva Edison, Frederick Winslow Taylor, Henry Ford, Claude Shannon, John von Neumann, Peter Drucker, Bill Gates, Steve Jobs, Larry Page and Sergey Brin, being capable of powerful entropic thinking, demonstrated great visions amongst many other people. Visionary people define some strategic objectives in their futures and then design *strategies* as means for achieving them in the real world, which is full of uncertainty and associated risks. Inertial thinking cannot do it since it does not contain time in its modeling structure; dynamic thinking cannot do it since processes are reversible and future and past are treated as interchangeable time zones; entropic thinking is the only thinking model able to conceive a clear future which is not interchangeable with the past. Even though all the three thinking models co-exist within our mind, thinking for the future requires that entropic thinking should be dominant over all the other thinking models. *Strategic thinking* and *designing strategies* cannot be successful unless we understand these thinking models and their roles in decision making.

Considering the future and the potential development of some present technologies, it is interesting to show that many people were not able to conceive any successful outcome of some new inventions even if some of them were cognoscenti about the field of business. Szczerba (2015) selected 15 "worst tech predictions" spanning the past 150 years, that didn't quite turn out as expected. We shall present some of them to demonstrate that, without an entropic thinking, there is no chance of understanding the future. William Preece, Head of the British Post Office, declared in 1876 about the invention of telephone: "The Americans have need of the telephone, but we do not. We have plenty of messenger boys". William Orton, the President of Western Union, declared in 1876 about the same invention that: "This 'telephone' has too many shortcomings to be seriously considered as a means of communication". Early predictions were that Henry Ford's horseless carriage was simply a fad. For instance, the President of the Michigan Savings Bank advised Horace Rackham, who was Henry Ford's lawyer, not to invest in the Ford Motor Company, arguing that "The horse is here to stay but the automobile is only a novelty—a fad". Darryl Zanuck, of 20th Century Fox, predicted in 1946 that television had no future: "Television won't be able to hold on to any market it captures after the first 6 months. People will soon get tired of staring at a plywood box every night". Approaching our present times, Robert Metcalfe, who was the founder of 3Com, predicted in 1995 that the "Internet will soon go spectacularly supernova and in 1996 catastrophically collapse". Also, it was very surprising that Steve Ballmer, Microsoft CEO, predicted in 2007 that "There's no chance that the iPhone is going to get any significant market share".

4.1.4 The Flying Time

We have already discussed about the need of using conceptual metaphors in understanding abstract concepts or in defining new ones in Chap. 1. Since perception of time is strongly related to the perception of space in our experience, our mind creates some*time-space* metaphors. In all of these metaphors the source domain contains *space* as being the known semantic field, while the target domain contains *time* as less known semantic field. Differences in the perception of time come mostly out of different interpretations for its *direction* and *motion*, interpretations generated by different cultures and life experiences (Boroditsky 2000; Grondin 2010; Santiago et al. 2007).

Casasanto and Jasmin (2012, p. 643) show that "in spoken English, time appears to flow along the sagittal axis (front/back): the future is ahead and the past is behind us. Here we show that when asked to gesture about past and future events deliberately, English speakers often use the sagittal axis, as language suggests they should". As a result of our experience in which space and time are integrated in a common referential framework, almost every aspect of time can be expressed in spatial words (Evans 2004; Lakoff and Johnson 1980, 1999; Núñez and Sweetser 2006). For instance, we say that durations can be *long* or *short*, and events can be moved *forward* or *pushed back*. Also, "deadlines lie *ahead of us* or *behind us*; we can *look forward* to our golden days or *look back* on our childhood. Time is metaphorized as a horizontal line extending indefinitely ahead of and behind the speaker" (Casasanto and Jasmin 2012, p. 644). Thus, if we imagine ourselves as being positioned as a reference point on a sagittal line, then the future will be in front of us and the past will be behind us. "The image schema acquired from forward movement of the body from one point to another probably constitutes the experiential basis for our spatialization of time along the back-front axis" (Santiago et al. 2007, p. 512). In this metaphorical setting, time is lined down like space in a motionless state and the observer or experiencer is moving along the sagittal line from the past—which is positioned behind us—toward the future—which is positioned in front of us. This is called the "ego-moving" perspective (Lakoff and Johnson 1999; Santiago et al. 2007). It is maybe the mostly used perspective in strategic management, since planning means to look ahead for structuring business activities. If we assume that the sagittal line constitutes the only possible direction for achieving the desired objectives, strategic planning becomes a linear process and it is based on the assumption that the environment is not changing or it is changing but in a predictable way. The ego-moving spatialization of time creates the illusion that future is just an extension of the present time and stimulates *deterministic thinking*. That is a limitation of strategic thinking, but the early theories of strategic management dealt with such a perspective in designing deliberate strategies (Mintzberg 2000; Mintzberg et al. 1998). The theories elaborated in the 1970s showed that strategic planning "did not create the strategy so much as pursue the strategy created by other means: it programmed the consequences of the given strategy, in terms of funds to raise, facilities to build, workers to hire, budgets to prepare, and so on. We can call this *deterministic planning*: specifying a determined course of action for the organization" (Mintzberg 2000, pp. 240–241).

The *ego-moving* perspective takes the model of driving along a known road: we know the part of the road which remains behind us and we look ahead toward the next destination. If the road is in an open space we can see far away in front of us to the next part of the road. Mapping that image onto the time domain, we get that our past is behind us because we have already went through it. Also, we consider the future in front of us because we see the objectives planned to achieve and the managerial way of deploying organizational resources in order to achieve those objectives. However, mapping the roadmap onto the planning process generates a certain inflexibility of planning: "In what may be a more controversial, but ultimately more important point, we wish to argue that planning itself breeds a basic inflexibility in organizations, and so a resistance to significant change" (Mintzberg 2000, p. 175). Thus, strategic planning is rather a conservative process which is in conflict with the basic idea that future is not an extrapolation of the present. As a consequence, deliberate strategies that lead to deterministic planning should be reconsidered by using other space-time perspectives.

The motion between observer and time is relative. Thus we can switch the setting and consider the observer stationary and the time flying. This is the *time-moving* perspective "with future events frontally approaching and passing him or her by toward the back" (Santiago et al. 2007, p. 512). In this perspective, when future comes to us we cannot see far ahead and do not have necessary time to plan ahead our activities. Since future comes frontally to us we must react in concordance with the planned objectives by delivering *emergent* strategies. If deliberate strategies imply a pro-active attitude and strategic thinking, emergent strategies imply a reactive attitude trying to stay within the framework conceived by the deliberate ones. Emergent strategies are more appropriate when the business environment becomes turbulent and the inflexibility of deliberate strategies cannot satisfy the changing environment anymore (Johnson et al. 2011; Mintzberg 2000). It is interesting to remark the concordance between the perception of the future coming to us and the countdown practice in the American culture. We will discuss the notions of emergent and deliberate strategies more thoroughly in the next chapters, but now we just wanted to highlight the strict connection between a certain perspective on strategies and the way time is seen.

In both the perspectives presented above, a future event is seen in front of the experiencer—assuming that he or she is facing the flow of events—whereas a past event is considered behind the experiencer. As Núñez and Sweetser remark (2006, p. 402), "all documented languages (with the exception to be discussed later) appear to share a spatial metaphor mapping future events onto spatial locations in front of Ego and past events onto locations behind Ego, rather than to the left of Ego and to the right of Ego, for example". The exception is Aymara language and culture. Aymara is an Amerindian language spoken by people living in the Andean highlands of western Bolivia, southeastern Peru, and northern Chile. "In Aymara, the basic word for FRONT (*nayra*, "eye/front/sight") is also a basic expression meaning PAST, and the basic word for BACK (*qhipa*, "back/behind") is a basic expression for FUTURE meaning" (Núñez and Sweetser 2006, p. 402). The logic of this metaphor is that the future is unknown and as a consequence its positioning should be behind us, while the past is known and thus it can be placed in front of us. This result has been obtained also by analyzing speech-accompanying gesture which is less conscious than language, and reflects much better this way of thinking.

Some researchers (Casasanto and Jasmin 2012; Ouellet et al. 2010; Santiago et al. 2007) focused their attention on the metaphors where time is flying along a lateral axis. These metaphors are related with the direction of reading and writing: from left to right, or from right to left. Tversky and her colleagues (1991) imagined a very simple but intuitive experiment with children and adults of English-speaking culture and Arab-speaking culture. Participants were requested to place stickers on a page to indicate where breakfast and dinner should appear relative to the lunch sticker, which was positioned in the middle of the page. "Whereas English speakers placed breakfast on the left and dinner on the right of lunch, Arabic speakers preferred the opposite arrangements, consistent with the direction of reading and writing in English and Arabic, and with the lateral organization of time on calendars in English- and Arabic-speaking cultures" (Casasanto and Jasmin 2012, p. 648). The lateral axis has been influenced also by the ordering the numerical system: in the European speaking cultures, numbers are placed from left to right in increasing order; in Arab speaking cultures it is the reverse. It is interesting to note that Arab children start using this right-left direction less frequently when they learn English.

"These findings suggest that the concept of time is mapped onto a horizontal axis running from left to right, as expected from the reading habits hypothesis, but do not rule out the possibility that these results are due to universal perceptual, motoric, or cerebral factors" (Ouellet et al. 2010, p. 309).

Since the lateral axis metaphor is a result of the way people read and write, planning is structured usually into sequences which follow the same axis. Thus, earlier events are positioned to the left and further events are positioned to the right. A good illustration can be also the Gantt chart when activities are arranged along this lateral left-right direction. "The proposed strategy then results in the habit of placing earlier events on the left mental space followed by later events being located more to the right" (Ouellet et al. 2010, p. 312). Integrating all these above experiments and empirical results makes us to conclude that conceiving a strategy is intimately related to our perception of time which is based on the space-time conceptual metaphors.

4.2 Dealing with Complexity

4.2.1 Linear Thinking

Understanding the future means understanding complexity. As we have learnt from the previous discussion, future is not a simple extension of the present with simple correlations between phenomena. Future is not simply revealed by means of a known algorithm, but it will come up with many new and complex events and phenomena. Dealing with complexity means finding practical ways of developing some mental models able to explain different levels of complexity. The simplest level is that of *linear thinking* which represents the most widespread approximation of complex problems. As Hastie and Dawes remark (2001, p. 52), "Our subjectively experienced world is dominated by approximately linear relationships". In the literature, there are two different perspectives concerning the meaning of linear thinking. For Groves et al. (2008) and Vance et al. (2007) *linear thinking* is identical with *rational thinking*, a perspective coming from the famous Cartesian dualism of mind and body (Russel 1972). Their model defines *linear thinking* style as "a preference for (1) attending to external tangible data and facts, and (2) processing this information through conscious logic and rational thinking to form knowledge, understanding, or a decision for guiding subsequent action" (Groves et al. 2008, p. 309).

The second perspective has been developed in Bratianu (2007, 2009), and Bratianu and Vasilache (2010), and is based on the mathematical concept of *linear space*. A mathematical space is considered to be linear if all linear combinations of its elements yield entities which are elements of this space. For instance, if 3 and 5 are elements of a liner space, then their product, which is 15, belongs to that space as well. Using linear spaces we can define linear equations and using linear equations we can model a linear process. From a very practical point of view, *any process whose output is proportional with its input is a linear process*. That means that any correlation based on proportionality between its input and output is a linear correlation and can be described by a linear equation. If there are several linear correlations and we combine them, the result will be also a linear correlation in concordance with the superposition principle. In reverse, a linear complicated problem can be decomposed into simpler problems, which can be solved individually. Then, the solutions obtained to these simple problems can be combined to yield the complete solution to the initial problem. That is valid only for linear problems and correlations. In the followings, we shall consider this second perspective based on linear spaces. Compared with the concept defined in Bratianu (2007, 2009) and based on linear spaces, the concept of linear thinking defined by Groves et al. (2008) looks like an extension to cover all rational reasoning processes. However, that extension has no scientific basis and covers many nonlinear phenomena. Knowledge and intellectual capital, defined in a rational way, should be in accordance with that definition of linear concepts, although they have a nonlinear nature (Bratianu 2015b; Davenport and Prusak 2000; Nonaka and Takeuchi 1995; Ricceri 2008; Roos et al. 2005).

Linear thinking is based on such linear cause-effect relationships, which represent actually cognitive approximations of more complex relationships and processes. In practice, linear thinking turns into *linear metrics* which contributed substantially to the linearization of our social life. Almost all systems to measure physical quantities are based on linear metrics. For instance, a thermometer uses the phenomenon of dilatation of mercury or other liquid, dilatation which is proportional with the value of the temperature field. Budgetary salaries are computed by using linear metrics and many universities in Europe use linear metrics in designing their curricula. We may say that these universities produce linear diplomas. The work productivity and many economic indicators are measured using linear metrics. Even in academic life, performance is judged by using linear thinking in many universities, which means to equate performance with the number of published papers. Schools and universities are evaluated and ranked on linear metrics. The temptation of using linear metrics can be seen also in many evaluation models designed to measure the organizational intellectual capital which is by definition nonlinear. According to Hastie and Dawes (2001, p. 61), "The *psychological principle* that might explain the predictive success of linear models is that people have a great deal of difficulty in attending to two or more non-comparable aspects of a stimulus or situation at once".

Linearity is based on summation of numbers or of some physical objects which means that their nature remains unchanged. There is no transformation. We are almost prisoners of linear thinking, since it is really difficult to escape from such a mental framework. "The simple linear model is surprisingly successful in many applications. We say surprisingly because many judges claim that their mental processes are much more complex than the linear summary equation would suggest, but empirically the equation does a remarkably good job of 'capturing' their judgment habit" (Hastie and Dawes 2001, p. 53). Strategy design has been based initially on linear extension of the present trends being supported by many econometric linear models. That is why many of them failed and could not achieve the expected results. Linear thinking cannot deal with complex nonlinear phenomena and that is why it should be replaced in elaborating strategies by nonlinear thinking models. That becomes a must for knowledge strategies since knowledge is strongly nonlinear.

4.2.2 Nonlinear Thinking

When a pot with water is put on a stove the heat transferred to the water increases its temperature linearly up to 100 °C. If the pot remains on the stove, although the water continues to receive heat its temperature remains constant at 100 °C and the boiling water starts transforming into steam. This phase transformation of water into steam is not a linear process anymore. It is nonlinear and for its description we must use nonlinear correlations and equations. The proportionality condition between the cause and the effect is replaced by other nonlinear conditions like logarithmic, exponential, polynomial, or even more complex ones. *Nonlinear thinking* is based on *nonlinear correlations and equations.* In nature as well as in society most phenomena are nonlinear. Nonlinearity is the rule and linearity is the exception. However, due to its complexity, nonlinearity is approximated with linearity in many fields and that explains why education contributes to the development of linear thinking. However, as Ohmae (1982, p. 13) remarks, "Phenomena and events in the real world do not always fit a linear model. Hence the most reliable means of dissecting a situation into its constituent parts and reassembling them in the desired pattern is not a step-by-step methodology such as systems analysis. Rather, it is that ultimate nonlinear thinking tool, the human brain".

Because the output of a given process is not correlated proportionally with the input to that process, we may face the situation when with a very small effort a very large effect can be obtained. This is called the *butterfly effect,* a metaphor used by Edward Lorenz when he was researching the weather dynamics at the famous Massachusetts Institute of Technology. He was using a deterministic system of equations to search for weather evolution in time. When Lorenz introduced the same initial conditions into the computer, the system displayed the same behavior. When he started the program with a slightly different initial condition, the outcome changed unexplainably. "Given a slightly different starting point, the weather should unfold in a slightly different way. A small numerical error was a small puff of wind—surely the small puffs faded or canceled each other out before they could change important, large-scale features of the weather. Yet in Lorenz's particular system of equations, small errors proved catastrophic" (Gleick 2008, p. 17).

We have been trained in schools to think that "what goes into any transactions or relationship or system must be directly related, in intensity and dimension, to what comes out" (Gladwell 2010, p. 11). However, many phenomena in the real life, instead of following the proportionality rule, manifest the *butterfly effect.* For instance, epidemics are a known example of geometric progression: when a virus spreads through population, it doubles at each new wave of disintegration

(Gladwell 2010). The same logic of the *butterfly effect* applies in the case of exponential developing phenomena like explosions. For instance, the nuclear explosion that happened at Chernobyl (Ukraine) on 26th April 1986 is a typical situation when people could not understood strong nonlinear phenomena due to their linear thinking. The nuclear reactors used at this power station had been designed based on an old technology with a low level of nuclear safety. During an experiment run on the nuclear reactor number 4, almost all content of the cooling water inside the reactor was at saturation temperature. A very small uncontrolled increase in the nuclear fission reactions resulted in a sufficient heat generation able to transform the water into steam. When a particle of water transforms into steam its specific volume increases hundreds of time. That small heat generation input made a huge quantity of water to be transformed almost instantaneously into steam increasing its volume immensely. The result was a steam explosion which destroyed the reactor vessel and threw huge quantities of nuclear fuels into the atmosphere. That explosion became very soon a nuclear catastrophe due to its fantastic force and the multiplying action of the winds. As Gladwell (2010, p. 11) remarks, "We need to prepare ourselves for the possibility that sometimes big changes follow from small events, and that sometimes these changes can happen very quickly".

Although industrial management created by Frederick Taylor has been based on many linear economic indicators, knowledge management needs nonlinear thinking since all the aspects of that process are strongly nonlinear. Knowledge, especially emotional and spiritual knowledge, intellectual capital, motivation, change, organizational culture and knowledge leadership cannot be managed by using linear thinking. They need a nonlinear approach. The same happens with *experience* which is the source of tacit knowledge (Bratianu 2015b; Nonaka and Takeuchi 1995). Many people equate experience with the number of years spent in front of a computer or performing same routines over and over again. But experience is not a linear aggregation of data and information; it is an integration and filtration of them such that the final result is much more powerful due to the processing of raw data and information. However, many people having interviews for new jobs are frequently asked "How many years of experience they have" in the domain of that job, i.e. experience is measured with the linear time metric. Due to the same linear thinking phenomenon, many people equate the quality of a decision with the time involved in making rational analyses, ignoring those decisions based on intuition which reflects a rich experience: "the task of making sense of ourselves and our behavior requires that we acknowledge there can be as much value in the blink of an eye as in months of rational analysis" (Gladwell 2005, p. 17). Nonlinear thinking requires a greater effort of discovering the underlining correlation between the cause and the effect but it delivers a more adequate solution to a complex problem.

Future cannot be linear since it is not a simple extension of the present. Future is just another time domain with complex events and phenomena which can be approached by using nonlinear thinking. Strategies should be conceived by using nonlinear thinking models and their implementation should be designed keeping in mind that organizational change is a complex nonlinear process. The linearity bias can be overcome only by changing our mental settings.

4.2.3 Systems Thinking

Our education is based mostly on thinking models composed of simple correlations with very few variables. Real processes and phenomena involve many variables and interactions such that any decomposition of the whole would change its nature and behavior. For such a situation, we have to develop *systems thinking* which considers a *system* as a base entity. A *system* is an assemble of interconnected elements designed to perform a given task. A system is not a simple collection of different objects put together, but an integral result of all these interconnections which yield synergy effects. "It is the connectedness of the component parts of the system with each other, and of the system as a whole with its environment, that is the central reason for order being maintained, and indeed created. Self-organizing systems all exchange energy with their environments and so fall into a class referred to as *open* systems" (Sherwood 2002, p. 15).

An outstanding contribution to the development of systems thinking has been made by Peter Senge (1999). According to his argumentation, "Today, systems thinking is needed more than ever because we are becoming overwhelmed by complexity. Perhaps for the first time in history, humankind has the capacity to create far more information than anyone can absorb, to foster far greater interdependency than anyone can manage, and to accelerate change far faster than anyone's ability to keep pace" (Senge 1999, p. 69). Systems thinking means to comprehend the whole system in its integrality and functionality without any decomposition. "Systems thinking is a discipline for seeing wholes. It is a framework for seeing interrelationships rather than things, for seeing patterns of change rather than static *snapshots*" (Senge 1999, p. 68). By decomposition, a system loses its integrality and thus changes its nature. That is true not only for living systems but also for organizations which are social systems and represents wholes.

Systems thinking is a conceptual framework which implies four basic disciplines: personal mastery, mental models, building shared vision, and team learning (Senge 1999). *Personal mastery* reflects our capacity of deepening our personal vision and spiritual understanding of a living company (De Geus 2002). Personal mastery is a cornerstone in developing a learning organization by enhancing knowledge sharing and organizational learning. *Mental models* are cognitive approximations of the real world based on assumptions, generalizations, correlations and experience processing which allow us to understand the world we live in. Mental models develop in time as a result of our education and cultural values we learn in family and society. In designing strategies for achieving a competitive advantage, mental models become essential in understanding business trends and creating business scenarios (De Geus 2002). *Building a shared vision* is one of the most important tasks of leadership. Without a shared vision managers cannot align their efforts in obtaining a competitive advantage. A company's shared vision cannot be imposed top-down; it must be develop in time through stimulating building trust, sharing knowledge, sharing organizational values and goals. *Team learning* is the engine of organizational learning and developing a learning organization. Team learning means for all the members of a certain team to share their rational, emotional and spiritual knowledge and to contribute in creating new adaptive and generic knowledge (Bratianu 2015b; Nonaka and Takeuchi 1995; Senge 1999).

Understanding the future and designing strategies cannot be effective without developing systems thinking due to complexity of new phenomena and processes any probable future brings with it. Systems thinking focus on the wholes and the interactions between their components. In this perspective, strategies appear as a learning process since understanding wholes is much more difficult than understanding simple correlations between two variables.

4.3 From a Single Certain Future to Multiple Probable Futures

4.3.1 Deterministic Thinking

Deterministic thinking is based on the assumption that events and phenomena must be *well-defined* and *well-determined* before they happen. Actually they happen due to our way of determining them. Their occurrence is certain. We also may say that chances for such events to happen are 100%. Consider for instance the train or the airplanes time tables. Their departure and arrival time is well defined, as well as their connections with other trains or airplanes. In real life there could be some delays or even cancelations due to weather conditions or some other technical problems but their time table remains a deterministic thinking model without any doubts. Deterministic thinking has been developed by scientists and engineers and reflects a deep human aspiration for stability and safety (Kahneman 2011; Taleb 2007). Scientists discovered laws able to describe and explain natural phenomena and to predict future developments. Consider for instance the law of energy conservation which says that energy cannot be created and cannot be destroyed; it can only be transformed from one form into another. It is well-formulated and very clear. There is no ambiguity or uncertainty involved in its formulation. Based on these scientific laws, engineers developed all kind of technological systems whose operation is well-determined and without any uncertainty. We have no doubt that turning the driving wheel toward right, the car will change its direction toward right. All technological systems must execute the tasks for which they have been created and nothing else. Thus all the procedures designed to operate technological systems represents also a product of a deterministic thinking.

In management, deterministic thinking is related to creating order and reducing the organizational entropy. Industrial management is well-known for its *command-and-control* philosophy which means order and discipline. Frederick Taylor and Henry Fayol—the pioneers of that kind of management approach —were engineers and they used in their managerial philosophy a mechanical model for organizations.

These models are still used by many managers, especially those who work in mass production economy and for whom deterministic thinking translated into regulations and procedures is the best way of managing people. As Morgan (1997, p. 13) observes correctly, in all of these situations "We talk about organizations as if they were machines, and as a consequence we tend to expect them to operate as machines: in a routinized, efficient, reliable, and predictable way". Thus, efficiency, reliability, and predictability constitute the major advantages of using deterministic thinking in management. However, deterministic thinking reduces dramatically the organizational entropy which reduces the potential of creativity and innovation. Also, deterministic thinking promoted by Max Weber in administration led to bureaucracy: "as a form of organization that emphasizes precision, speed, clarity, regularity, reliability, and efficiency achieved through the creation of a fixed division of tasks, hierarchical supervision, and detailed rules and regulations" (Morgan 1997, p. 17).

People with deterministic thinking consider that future is certain representing an extrapolation of the present time. Even if there will be some changes they can be anticipated based on the present trends and thus there will be no known unknowns in designing deliberate strategies. Strategic planning will be an extension of the operational planning which means to use the same logic and metrics in designing it. However, future cannot be fully anticipated and strategic planning based on deterministic thinking cannot adapt to a changeable future. That is why strategic planning as it has been conceived initially could not pass the test of time (Mintzberg 2000).

4.3.2 Probabilistic Thinking

Future does not exist simultaneously with present time. It will develop as a new time zone connected to the present but not in a rigid way. Future will contain events and phenomena which cannot be completely anticipated. Thus, future comes with an associated uncertainty which induces the need for a different way of thinking. We call it *probabilistic thinking* because it deals with probable events and phenomena (Kahneman 2011; Syrett and Devine 2012; Taleb 2007). In nature and society events and phenomena happen randomly and we live practically in a continuous uncertainty. Although our education emphasized deterministic thinking, in the real life uncertainty is the rule and certainty is the exception. That is why people have usually a tendency to avoid uncertainty and deal with thinks which are well-determined. A good example of a probabilistic model is the weather forecast. If the forecast is done for the next day, chances are to have a fairly good prediction. If the forecast is done for the next week, the accuracy of that prediction is questionable. The farther in time we want to know the weather forecast, the less we will be able to formulate it due to increasing uncertainty. Probabilistic thinking is closely related to risk management and decision making in conditions of uncertainty. "The essence of risk management lies in maximizing the areas where we have some control over the outcome while minimizing the areas where we have absolutely no control over the outcome and the linkage between effect and cause is hidden from us" (Bernstein 1998, p. 197).

Uncertainty is not an objective characteristic about a real phenomenon. It is about the relationship between the knower and a given context in the present or in the future. "Statements of uncertainty are personalistic, they belong to the person making them and express a relationship between that person and the real world about which a statement is being made. In particular, they are not objective in the sense that they express a property that is the same for all of us" (Lindley 2006, p. 1). However, uncertainty is generated by the absence of knowledge in all its aspects. Reflecting on the multiple dimensions of uncertainty Spender (2014) considers of being important the followings: (a) ignorance, which reflects the absence of needed information and knowledge about a future event or phenomenon; (b) incommensurability, which reflects the absence of a common metric for what it misses; and (c) indeterminacy, which reflect the unknown way people will act in that future. "Thus, in addition to *ignorance* and *incommensurability*, people generate *indeterminacy* as they respond actively to each other and change in ways contingent on that action. The firm's context becomes interactive, dynamic in a different way" (Spender 2014, p. 11).

When switching from a generic uncertainty reflecting a state of knowledge absence about a given context to a specific uncertainty which focuses on a particular aspect of that context, or a certain event, some authors refer to belief which "expresses a relationship between you and the world, in particular between you and an event in that world. The word that will be used to measure the strength of your belief is *probability*" (Lindley 2006, p. 12). Thus, we can measure uncertainty generated by some random events by using the concept of *probability* which is defined as a ratio of the favorable cases to the whole number of cases possible. Knight (2006) distinguishes different contexts of uncertainty and probability assessment:

- *A priori probability*. That is the mathematical concept defined for a class of homogeneous random events.
- *Statistical probability*. That is based on empirical evaluation of the frequency of occurrence of a certain event in similar contexts. "The main distinguishing characteristic of this type is that it rests on an empirical classification of instances" (Knight 2006, p. 225).
- *Estimates*. We use them when the absence of knowledge makes the application of the probability theory almost impossible. "This form of probability is involved in the greatest logical difficulties of all, and no very satisfactory discussion of it can be given" (Knight 2006, p. 225).

This uncertainty evolution evolves from "known knowns" about the random events to "unknown unknowns" which makes probability operational only for the first two situations. For estimates decision makers should use some other techniques like foresights and scenario building (De Ruijter and Alkema 2014; Murgatroyd 2015). We shall discuss about them in the next section.

Prom a psychological perspective, Amos Tversky and Daniel Kahneman researched the influence of uncertainty upon the rational decision making and developed the *Prospect Theory* (Bernstein 1998; Kahneman 2011). "Prospect Theory discovered behavior patterns that had never been recognized by the proponents of rational decision-making. Kahneman and Tversky ascribe these patterns to two human shortcomings. First, emotion often destroys the self-control that is essential to rational decision making. Second, people are often unable to understand fully what they are dealing with" (Bernstein 1998, p. 271). The Prospect Theory shows that a decisive role in making decisions is played by the asymmetry between the way we make decisions involving gains and decisions involving losses. In one of their experiments (Bernstein 1998, pp. 272–273) they asked the subjects to choose between the following alternatives: (a) an 80% chance of winning $4000 and a 20% chance of winning nothing, versus (b) a 100% chance of receiving $3000. Although the first alternative has a higher mathematical expectation, 80% of all subjects chose the second alternative. These subjects demonstrated a preference for a risk-averse behavior. Then, the experiment has been reversed in negative terms: (a) an 80% chance of losing $4000 and a 20% chance of breaking even, versus (b) a 100% chance of losing $3000. Now, 92% of all respondents chose the first alternative. When the choice involves losses, we are risk-seekers, not risk-averse.

Probabilistic thinking brings new perspectives in understanding future as a generator of uncertainty and in how to deal with probability of events and phenomena as a function of their occurrence. Designing strategies for achieving a competitive advantage means to understand how rational decision making is influenced by the asymmetry of information and the natural aversion toward risk taking.

4.3.3 Probable Futures

In all our discussions so far, we considered only one future with events and phenomena having different behavior from certain to uncertain. In deterministic thinking, we consider that all events and phenomena are certain and they will be part of a predictable future which is an extension of the present time. For such a future, managers extend their operational planning into a strategic planning, by changing only the time scale. In probabilistic thinking, we change the paradigm of the future by considering that it is populated with events and phenomena which have a random behavior. Understanding the future, in this case, means to describe the probability distributions of them if their occurrences are homogeneous, or to compute statistically their likelihood in the real life. When we face with rare events for which the absence of knowledge is severe, we can make estimates based on similar cases.

Although we can live in only one future, speaking in the probabilistic perspective there could be more probable futures. Each future contains probable events and phenomena which have some common characteristics. Understanding the future, in this case, means to change the paradigm of defining a future. Multiple probable futures can be imagined if we develop different probable scenarios based on different assumptions or hypotheses. The logic of such a paradigm comes from the fact that in the real life the business environment changes sometimes in a disruptive way due to a new *disruptive technology* or a *disruptive innovation*. According to Christensen (2003, p. xviii), "Disruptive technologies bring to a market a very different value proposition than had been available previously". In the paradigm of a single future we use *prediction* to measure the chances of different events to happen. In the paradigm of multiple probable futures we replace prediction with *foresight* and *scenarios* (De Ruijter and Alkem 2014; Murgatroyd 2015; van der Laan and Yap 2016). Foresight means to discover and understand the driving forces of changing the business landscape by processing real-time and historical structured and unstructured data. Foresight is a complex process involving four phases (Murgatroyd 2015, p. 11):

- "Examining and projecting current trends and issues.
- Considering potential events or tipping points and 'wild cards'.
- Developing possible futures.
- Choosing among the different futures available the most preferred future."

Looking from the present moment to the infinity of the timeline we can distinguish four domains of the future: (a) preferable future; (b) probable futures; (c) possible futures; and (d) plausible futures (Murgatroyd 2015, p. 73). The key to the new paradigm is to identify early signals and significant trends of change and to imagine toward what domain of the future will target those trends. Then, to come back to the present time and develop scenarios in concordance with those trends and anticipated changes. That means that in a turbulent business environment it is not vital to predict *what will happen*, but to explore *the different things which could happen*. "Taking into account various possible futures based on one or more models enables an organization to decrease the reaction time needed when a new development actually occurs. This changed scenario thinking from a method to predict the future into a method to explore possible alternative futures and to think through what the organization could do in these possible situations" (De Ruijter and Alkema 2014, p. 57). We may consider that scenario thinking doesn't aim at making better predictions about the future, but at helping organizations to be better prepared for the coming future.

4.3.4 Future as Chaos

The diversity of thinking models and approaches presented so far for understanding the future or the probable futures can be integrated into the most complex perspective of thinking—the chaos theory. *Chaos* is a multidimensional domain of the new science which integrates theories about dynamical systems, strange attractors, fractals, bifurcations, intermittencies and periodicities, folded-towel diffeomorphisms and smooth noodle maps (Bird 2003; Gleick 2008). As Gleick

(2008, p. 5) underlines, "These are the new elements of motion, just as, in traditional physics, quarks and gluons are the new elements of matter. To some physicists chaos is a science of process rather than state, of becoming rather than being". We may extend that last idea of the future since it is a becoming process and not a state. Future does not evolve according to the Newtonian laws from the known present and past but from the complexity of the random interactions between an infinite of potential events and phenomena. Although it looks almost impossible in the perspective of classical science, chaos researchers look for some order in that infinite disorder. "At first sight, chaotic motion might seem to be the complete disorder that the term traditionally implied. But such systems, although their behavior seems complicated, have in them an underlying pattern that can be used to predict and sometimes even control them" (Bird 2003, p. 63).

A certain pattern of order that attracts in time the behavior of a dynamic system has been coined *strange attractor*. "A strange attractor displays a recognizable pattern in space or over time but that pattern is irregular. In other words, strange attractors are paradoxically regular and irregular, stable and unstable, at the same time" (Stacey et al. 2000, p. 87). If we consider an organization a dynamical system and a strategy as a driving force toward the future, then that future will appear as an *attractor* which represents a pattern of random interactions between many unknown and unpredictable events and phenomena. Although that future cannot be constructed in a rational and predictable way like in classical strategic planning, it can be unfolded through continuous and discontinuous interactions between people, and between people and random events. A *strange attractor* may be interpreted as a *probable future* which can be imagined as a realization of one of the multiple *probable scenarios* we developed. In that case the organizational behavior should be modeled by recursively applied nonlinear equations, just as systems dynamics does. When an organization follows a strange attractor "its behavior is predictable at global, macro levels of description, but only in qualitative terms. At the specific micro level, predictability is confined to short-term local occurrences, leaving the specific long-term trajectory unpredictable due to the inability of humans to measure with infinite accuracy" (Stacey et al. 2000, p. 89).

Going beyond deterministic thinking that reflects causality in a certain world, and probabilistic thinking that reflects random behavior in a world of uncertainty, chaos theory suggests an integrative approach where local order described by nonlinear recursive equations is hidden in a whole of disorder which cannot be described by any equation. Future as chaos suggests a combined approach of using some deterministic models for elaborating the deliberate components of a strategy with the creation of probable scenarios for preparing the managers for the emergent components of that strategy.

4.4 Conclusion

Strategizing means finding solutions for complex problems in absence of knowledge and for a time domain which is dominated by uncertainty. Thus, the perception of the future constitutes a necessary prerequisite in elaborating and implementing strategies. The present chapter presents a structured spectrum of different thinking models, able to describe and explain some of the main features of a hypothetical future along three dimensions: time, complexity and uncertainty.

The simplest model on the *time dimension* is inertial thinking. This model does not contain time as a variable which makes its action timeless. That means that along the timeline of any organization there is no past, present and future. It is only a continuous present. Inertial thinking opposes to any change since change implies time. Thus, inertial thinking is not an adequate model for thinking for the future and building up strategies. The next thinking model we analyzed contains time but only from a quantitative point of view. Since dynamic thinking is based on reversible processes there is no clear order between past, present and future. That is because the model is based on theoretical processes and involves only the quantitative aspect of time. Entropic thinking is the most complex thinking model along the time axis. It incorporates time both quantitatively and qualitatively and it is based on real irreversible processes. The entropic model contains past, present and future in that order. Time is unidirectional from the past toward the future. Since future is characterized by absence of knowledge we need to develop strategies to deal with that situation and to create premises for achieving a competitive advantage. In the next part of the section, we presented some interesting aspects of thinking metaphorically about the future and its perception with respect to ego positioning.

Along the *complexity dimension* we presented linear and nonlinear thinking, as well as system thinking. In linear thinking the outcome is proportional to the input, such that the causal correlation between input and output variables in any process is represented by a linear equation. Since people always look for simple solutions to complex problems linear thinking is a dominant mental model in society. Linear thinking is used in measuring physical properties of different objects and in measuring time. However, the future cannot be conceived as a proportional outcome of the present and linear strategies for such a future would be failures. Nonlinear thinking constitutes a much better representation of the real events and phenomena in our mind. Future outcomes of the present actions will be for sure nonlinear. Going beyond the pure correlation of two variables, system thinking represents a more complex approach to real phenomena in nature and society. System thinking should be used for designing strategies for a desirable future.

The *uncertainty dimension* places us very close to understanding real future. Here, we should distinguish between thinking models designed to deal with a single certain future and thinking models able to deal with multiple probable futures. Deterministic thinking reflects the Newtonian logic and is a dominant model of thinking in science and technology. It is based on certain laws which yield same results to similar problems when initial and boundary conditions are the same. However, future cannot be well-determined and precisely described by using these deterministic laws. Probabilistic thinking is based on the idea that the world we are living in is characterized by uncertainty. Certainty represents only limiting situations and the exception to the rule. Probabilistic thinking considers events and phenomena having a random behavior, and scientists developed complex theories of probabilities to deal with such situations. Future, in this case, appears to be a world of random events which cannot be predicted accurately and for any decision making there are some risks associated. Strategies will maximize, in this perspective, the opportunities and will decrease the risks. Profits are always correlated with the risks involved in the investments. By contrast to all of these previous thinking models for which there is only one future, new approaches based on scenarios reveal the possibility of considering a multitude of probable futures. Developing such kind of thinking for managers is important to prepare them for those disruptive technologies and innovations which change completely the business future and market competition. Finally, the chaos theories developed in the last decades come to show that both deterministic thinking and probabilistic thinking can be integrated such that we will have a chaotic future. Successful strategies will be those able to generate a dynamic organizational behavior targeting to reach a strange attractor.

References

Atkins, P. (2010). *The laws of thermodynamics: A very short introduction*. Oxford: Oxford University Press.

Ben-Naim, A. (2012). *Entropy and the second law*. London: World Scientific.

Bernstein, P. L. (1998). *Against the gods: The remarkable story of risk*. New York: Wiley.

Bird, R. J. (2003). *Chaos and life: Complexity and order in evolution and thought*. New York: Columbia University Press.

Boroditsky, L. (2000). Metaphoric structuring: Understanding time through spatial metaphors. *Cognition, 75*, 1–28.

Bratianu, C. (2007). Thinking patterns and knowledge dynamics. In *Proceedings of the 8th European conference on knowledge management* (Vol. 1, pp. 152–157). Reading: Academic Conferences Limited.

Bratianu, C. (2009, April 28–29). *The frontier of linearity in the intellectual capital metaphor* (pp. 97–103). Proceedings of the European Conference on Intellectual Capital, Inholland University of Applied Sciences, Haarlem, The Netherlands.

Bratianu, C. (2015a). Developing strategic thinking in business education. *Management Dynamics in the Knowledge Economy, 3*(3), 409–429.

Bratianu, C. (2015b). *Organizational knowledge dynamics: Managing knowledge creation, acquisition, sharing, and transformation*. Hershey: IGI Global.

Bratianu, C., & Vasilache, S. (2010). A factorial analysis of the managerial linear thinking model. *International Journal of Innovation and Learning, 8*(4), 393–407.

Casasanto, D., & Jasmin, K. (2012). The hands of time: Temporal gestures in English speakers. *Cognitive Linguistics, 23*(4), 643–674.

Christensen, C. M. (2003). *The innovator's dilemma*. New York: HarperCollins.

Davenport, T. H., & Prusak, L. (2000). *Working knowledge: How organizations manage what they know*. Boston: Harvard Business School.

De Geus, A. (2002). *The living company: Growth, learning and longevity in business*. London: Nicholas Brealey.

De Ruijter, P., & Alkema, H. (2014). *Scenario based strategy: Navigate the future*. Farnham: Gower Publishing Limited.

Evans, V. (2004). *The structure of time: Language, meaning and temporal cognition*. Amsterdam: John Benjaminis.

Fauconnier, G., & Turner, M. (2002). *The way we think: Conceptual blending and the mind's hidden complexities*. New York: Basic Books.

Frith, C. (2007). *Making up the mind: How the brain creates our mental world*. Oxford: Blackwell.

Georgescu-Roegen, N. (1999). *The entropy law and the economic process*. Cambridge: Harvard University Press.

Gladwell, M. (2005). *Blink: The power of thinking without thinking*. New York: Back Bay Books.

Gladwell, M. (2010). *The tipping point: How little things can make a big difference*. London: Abacus.

Gleick, J. (2008). *Chaos: Making a new science*. New York: Penguin Books.

Grondin, S. (2010). Timing and time perception: A review of recent behavioral and neuroscience findings and theoretical directions. *Attention, Perceptions, & Psychophysics, 72*(3), 561–582.

Groves, K. S., Vance, C. M., & Paik, Y. (2008). Linking linear/nonlinear thinking style balance and managerial ethical decision making. *Journal of Business Ethics, 80*, 305–325.

Handscombe, R. D., & Patterson, E. A. (2004). *The entropy vector*. London: World Scientific.

Hastie, R., & Dawes, R. M. (2001). *Rational choice in an uncertain world: The psychology of judgment and decision making*. Thousand Oaks: Sage.

Johnson, G., Whittington, R., & Scholes, K. (2011). *Exploring strategy: Text & cases* (9th ed.). London: Prentice Hall.

Kahneman, D. (2011). *Thinking fast and slow*. New York: Farrar, Straus and Giroux.

Knight, F. H. (2006). *Risk, uncertainty and profit*. Mineola: Dover Publications.

Kotter, P. J. (1996). *Leading change*. Boston: Harvard Business School Press.

Kotter, P. J. (2008). *A sense of urgency*. Boston: Harvard Business School Press.

Kotter, P. J., & Cohen, D. S. (2002). *The heart of change: Real-life stories of how people change their organizations*. Boston: Harvard Business School Press.

Lakoff, G., & Johnson, M. (1980). *Metaphors we live by*. Chicago: University of Chicago.

Lakoff, G., & Johnson, M. (1999). *Philosophy in the flesh: The embodied mind and its challenge to the western thought*. New York: Basic Books.

Lindley, D. V. (2006). *Understanding uncertainty*. New York: Wiley-Interscience.

Mintzberg, H. (2000). *The rise and fall of strategic planning*. London: Prentice Hall.

Mintzberg, H., Ahlstrand, B., & Lampel, J. (1998). *Strategy safari: The complete guide through the wilds of strategic management*. London: Prentice Hall.

Morgan, G. (1997). *Images of organization*. Thousand Oaks: Sage.

Murgatroyd, S. (2015). *How to rethink the future: Making use of strategic foresight*. Alberta: The Collaborative Media Group.

Nisbett, R. E. (2003). *The geography of thought: How Asians and Westerners think differently...and why*. New York: Free Press.

Nonaka, I., & Takeuchi, H. (1995). *The knowledge-creating company: How Japanese companies create the dynamics of innovation*. Oxford: Oxford University Press.

Núñez, R., & Sweetser, E. (2006). With the future behind them: Convergent evidence from Aymara language and gesture in the crosslinguistic comparison of spatial construals of time. *Cognitive Science, 30*, 401–450.

Ohmae, K. (1982). *The mind of the strategist: The art of Japanese business*. New York: McGraw-Hill.

Ouellet, M., Santiago, H., Israeli, Z., & Gabay, S. (2010). Is the future the right time? *Experimental Psychology, 57*(4), 308–314.

Ricceri, F. (2008). *Intellectual capital and knowledge management: Strategic management of knowledge resources*. London: Routledge.

Roos, G., Pike, A., & Fernström, L. (2005). *Managing intellectual capital in practice*. Amsterdam: Elsevier.

Rumelt, R. (2012). *Good strategy, bad strategy: The difference and why it matters*. London: Profile Books.
Russel, B. (1972). *A history of western philosophy*. New York: Simon & Schuster.
Santiago, J., Lupiañez, J., Pérez, E., & Funes, M. J. (2007). Time (also) flies from left to right. *Psychonomic Bulletin & Review, 14*(3), 512–516.
Senge, P. (1999). *The fifth discipline: The art and practice of the learning organization*. London: Random House.
Sherwood, D. (2002). *Seeing the forest for the trees: A manager's guide to applying system thinking*. London: Nicholas Brealey.
Spender, J. C. (2014). *Business strategy: Managing uncertainty, opportunity, & enterprise*. Oxford: Oxford University Press.
Stacey, R. D., Griffin, D., & Shaw, P. (2000). *Complexity and management: Fad or radical challenge to systems thinking?* London: Routledge.
Syrett, M., & Devine, M. (2012). *Managing uncertainty: Strategies for surviving and thriving in turbulent times*. London: Profile Books.
Szczerba, R. J. (2015). 15 worst tech predictions of all time (online). Retrieved from http://www.forbes.com/sites/robertszczerba/2015/01/05/15-worst-tech-predictions-of-all-time/
Taleb, N. N. (2007). *The black swan: The impact of the highly improbable*. London: Penguin Books.
Tversky, B., Kugelmass, S., & Winter, A. (1991). Cross-cultural and developmental trends in graphic productions. *Cognitive Psychology, 23*(4), 515–557.
Van der Laap, L., & Yap, J. (2016). *Foresight & strategy in the Asia Pacific region: Practice and theory to build enterprises of the future*. Berlin: Springer.
Vance, C. M., Groves, K. S., Paik, Y., & Kindler, H. (2007). Understanding and measuring linear-nonlinear thinking style for enhanced management education and professional practice. *Academy of Management Learning and Education, 6*(2), 167–185.

References

Knowledge Strategies 5

This chapter treats one of the central topics of the book: the notion of knowledge strategy. It is a concept that has started to become popular in the managerial literature only recently, and mainly due to the upsurge of the idea of knowledge economy and the diffusion of knowledge management (KM) programs. As we have seen in previous chapters, knowledge has always been a key topic for strategic formulation. However, the notion of knowledge strategy means more, because it suggests that a company should adopt a strategy to manage its knowledge assets. So, in addition to planning production and delivery of products and services, deciding goals regarding profits and markets, expressing objectives about competitive positioning, a knowledge strategy represents the effort to plan activities of KM and, more generally, all resources and processes that, in a company, are devoted to developing knowledge and competences of people, boosting learning processes, facilitating storage, sharing and reuse. As the chapter will show, the definition of knowledge strategy is however difficult, and it is still necessary to clarify its contents and boundaries. Also, there is some confusion with other terms used in the literature (such as KM strategy or knowledge management strategy). In addition, there is the need to clarify if a knowledge strategy is just a part—or a derivation—of classic strategic formulation of companies, or if it has a special and distinct place that also deserves specific approaches and methods.

5.1 An Important Notion but a Difficult Definition

Compared to that of strategy typically used in business, the notion of knowledge strategy is relatively recent, but its conceptual and practical importance is increasing. As we illustrated in the previous chapters, the strict connection between "strategy" and "knowledge", and the recognized importance of knowledge as a strategic factor in companies have inspired researchers to provide a definition of knowledge strategy. This is now a deliberate object of new models for helping managerial analysis and practice. The recognition of knowledge as a strategic

E. Bolisani, C. Bratianu, *Emergent Knowledge Strategies*, Knowledge Management and Organizational Learning 4, DOI 10.1007/978-3-319-60657-6_5

resource (Spender and Grant 1996) implies that the inclusion of knowledge strategy into the broader framework of strategic thinking must be carefully taken into account.

Particularly, it may be argued that having an appropriate knowledge strategy becomes essential for companies to improve and use their knowledge and intellectual assets to support competitiveness (Zack 2003). First of all, this is vital for companies whose business is mostly based on knowledge, which is especially the case of the so-called Knowledge-Intensive Business Services (KIBS—see e.g. Miles 2005): indeed, there is a growing number of companies whose main production output and outcome consists of knowledge, directly "sold" and delivered to customers or embedded into sophisticated services and even products. However, generally speaking, formulating appropriate plans to develop and manage knowledge is vital for all companies, including traditional manufacturers (Zack 1999a). In short, the formulation of a knowledge strategy should be strictly related, or it is better to say, aligned with the firm's general strategy (Eisenhardt and Santos 2002; Mládková 2014).

Another important point to recall is the upsurge of KM programs. These programs are increasingly popular in companies, and clearly require some form of planning—i.e. formulating KM goals, selecting relevant knowledge sources or resources, designing KM processes, appropriate structures, and implementation activities, etc. The notion of a strategy for KM, intended as long-term planning of organizational, managerial and technical resources that a company adopts for its KM programs, has progressively appeared in the literature (Holsapple and Jones 2007) and has increasingly attracted the attention of researchers (Coakes et al. 2010). A confirmation can be easily found in a recent analysis of the top-cited keywords used by authors who published in a major KM journal, *Knowledge Management Research & Practice* (Walter and Ribière 2013): the term "KM strategy" occupies the seventh position among the top 40 self-reported keywords, and "strategic KM" is the second most important focus of KM research. But a similar analysis can be done pretty easily, by just typing the key-phrase "knowledge strategy" into Google: more than 170,000 web pages can be retrieved, and more than 14,000 references in Google Scholar, the database specializing in scientific papers (these numbers were obtained in January 2017).

Also, empirical evidence shows that the conditions for the successful management of knowledge and intellectual assets can change from a company to another, in relation to the internal organization and the strategic directions, and to the external environment. Heterogeneity calls for approaches that fit the specific situation of companies, which may adopt different courses of action from one another.

But what is a knowledge strategy? Can we define this notion in a formal or at least a non-ambiguous way? This point is crucial, but problematic. If, as we have seen, it is difficult to define the notion of strategy in business, the case of knowledge strategy is the same, and even worse, considering that the notion itself of knowledge and its use in business are still ambiguous.

At present, there is still no absolute consensus on what a *knowledge strategy* is (Denford and Chan 2011; Donate and Canales 2012). There is, indeed, some agreement about its essential traits. An often cited definition is Zack's (1999a), who sees knowledge strategy as the *overall approach that an organization takes to align its knowledge resources and capabilities to the intellectual requirements of its business strategy*. Similarly, for von Krogh et al. (2001), the ultimate purpose of a knowledge strategy is the *application of "knowledge processes" to an existing or new knowledge domain to achieve a strategic goal*. Bierly and Chakrabarti (1996) see a knowledge strategy as the *set of choices and plans that regard the firm's knowledge base*, which affects its competitive capability. For Kasten (2007), a knowledge strategy can be referred to as the *general guidelines that shape the organization's capability to manipulate its cognitive resources*, with the ultimate goal, as Holsapple and Jones (2007) clarify, of *making the best use of these assets for competitive advantage*.

To summarize, by analyzing the notions proposed here, there are some core keywords that may represent the concept of knowledge strategy (Fig. 5.1). First, its objects are the *knowledge or cognitive resources* of a company (i.e. what is known and *understood* by employees and managers; what is embedded in files, documents, procedures, routines, projects, artifacts; what is represented by patents, licenses, etc.). Second, a knowledge strategy relates to *guidelines* and to *practical application of processes*, to make the best out of existing or new knowledge domains: this implies that a knowledge strategy should result in *plans to manage existing knowledge or creating new one*. Third, the ultimate goal is *achieving strategic or competitive advantage* by means of these knowledge resources. Fourth, developing and implementing knowledge strategies are about not only knowledge as a strategic resources, but also about *developing adequate capabilities to process information and knowledge* in concordance with the strategic objectives. A good example here is the emerging area of business analytics and big data analysis. Business analytics means to create dynamic databases containing all sorts of information concerning consumers' behavior, to develop intelligent software to process all this data and information, and to develop an internal capability for all employees to use those analytic applications. Indeed, as Davenport and Harris (2007, p. 7) underline, *analytics* means "extensive use of data, statistical and quantitative analysis, explanatory and predictive models, and fact-based management to drive decisions and

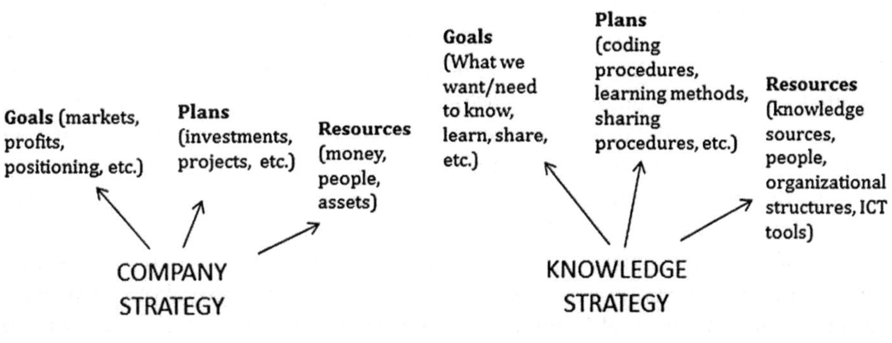

Fig. 5.1 Basic contents of a company strategy and a knowledge strategy

actions". So, the field of analytics develops as a new *cognitive capability* of firms. However, this is not purely processing data by means of "intelligent" systems, nor the exploitation of a "group of experts" or a special department of the firm. It is, instead, a capability that should be embodied in all knowledge workers. As Davenport and Harris (2007, p. 16) emphasize, "Analytical competitors will have to conduct experiments in many aspects of their businesses and learn from each of them. In order for quantitative decisions to be implemented effectively, analysis will have to be a broad capability of employees, rather than the province of a few 'rocket scientists' with quantitative expertise". In short, a knowledge strategy is about how to organize the company and its employees in a way that they can learn fast and become capable of analyzing information more effectively and transforming it into useful knowledge.

In this chapter, we will also explore the analogies and differences of the concept of knowledge strategy compared with the typical view of strategy in business that we have described in Chap. 3. For sake of simplicity, here we will use the term *company strategy* referring to the traditional notion of strategy in the business environment, although we are aware that other terms as *corporate strategy*, *competitive strategy*, *business strategy* are also used, sometimes as synonyms, some other times with partly different meanings. This distinction will be clarified better in the next sections.

Indeed, as we have examined in Chap. 3, the definition of company strategy and even the possibility to formulate it rationally, have been long debated and questioned (we will discuss this again in the rest of the book). Consequently, we may say that the definition of knowledge strategy still suffers from the same ambiguity.

However, even though the notions of strategy should be always contextualized to the different conditions where they are applied, we may say that when a company attempts to formulate a strategy, even in times characterized by turbulence and uncertainty, company executives need some practical references and ways of working, for example the compilation of a *business plan*. Still, in the most recent management textbooks, "writing a formal business plan is widely regarded as one of the most important aspects of strategic planning" (Kraus 2007, p. 76) for any kind of firms. This means that the idea of strategy in companies implies a reflection on *what we would like to achieve, where we would like to go, how, and when.* Maybe this reflection is partial, incomplete, not always rational, and temporary, but it is however an attempt to draw a *map of the territory* based on what we may know, and that we will use to guide our decisions until we find something better. So, in this section we will shortly analyze the generic *contents* of a company strategy, and we will also make a comparison with those that a *knowledge strategy* should have. In other words, we will characterize knowledge strategies by developing the analogies and differences with the more familiar idea of company strategy.

First of all, in the definition that we present here, a knowledge strategy appears to be like a company strategy but where the object is specifically about knowledge resources and intellectual capital. This means, for instance, that a knowledge

Table 5.1 Knowledge strategy: some examples of key issues

Issue	Main questions or topics	Key topics
Future knowledge of a company	What knowledge shall we need to compete in the future? Is what we already know sufficient, or do we need more? If so: how fast shall we need to learn?	Knowledge for organizational performance Knowledge for innovation
Kinds of knowledge required	What are the points of strength and weakness of the company? What is essential to know for the future business?	Technical, managerial, market knowledge Tacit vs. Explicit knowledge
Knowledge protection	Should a company protect the knowledge created internally? How to combine it with the necessity of networking with external partners?	Traditional intellectual property right protection vs. human-based knowledge protection Policies of knowledge disclosure
Culture of knowledge sharing	Is it important to facilitate knowledge sharing between employees in a company?	Interactions only along hierarchical lines Knowledge sharing across the organization Overcoming knowledge islands

strategy is typically about providing appropriate answers to specific issues (Table 5.1).

A first point regards a first group of key questions: what knowledge shall we need to acquire for competing in the future? Is what we already know sufficient, or do we need more? If so: how fast shall we need to learn? These questions are related to two main assumptions that lay in the background of a knowledge strategy (López-Nicolás and Meroño-Cerdán 2011; Forcadell and Guadamillas 2002): first, that the more effectively/efficiently a company can manage its knowledge resources, the higher are organizational performances; second, that the knowledge a company possesses, is capable to handle, or to learn, quickly influences and sustains the flow of innovations that it can produce. So, a knowledge strategy defines what the company should know for its future competition, and/or how it must acquire the capability to learn fast.

A second (and more specific) point regards the *kinds of knowledge* that a company should be prepared to handle for the future business: what are our points of strength and, conversely, what should we improve in terms of knowledge? Here, the literature has developed useful classifications that also inspire different *strategic options* to shape a company's capability to compete and innovate. For example, various studies have emphasized the importance of different kinds of knowledge in companies, i.e.: scientific knowledge (referring to scientific laws and experiments regarding natural phenomena), technical knowledge (i.e. application of scientific knowledge to practical purposes), market knowledge (namely, knowledge of demand, markets, clients), managerial knowledge (referring to capabilities to arrange and organize structures and processes), etc. Depending on the nature of companies, of their competitive environment, business, and future visions of executives, one type of knowledge or the other may tend to prevail or become more important (Hansen et al. 1999; Johannessen et al. 1999; Cillo 2005; De Luca and Atuahene-Gima 2007; Zhou and Li 2012). This also defines the possible choices regarding the knowledge strategy. Another very popular classification in KM, that between tacit or explicit knowledge, can also be of use to select different strategic formulations, in relation to the kind of knowledge that is considered more important in the particular company (Schulz and Jobe 2001; Senker 1995; Hansen et al. 1999).

A third issue regards the strategic orientation of a company towards the protection of the knowledge that is possessed or produced by the single individuals or the entire organizations. Considering the growing importance ascribed to innovation and fast go-to-market approaches, many companies make great efforts to protect the knowledge they create internally and, more generally, the investments in new knowledge creation. This leads to different strategies of knowledge protection that companies may adopt depending on their specific situation and goals; for example, traditional legal measures of Intellectual Property Right (McManis 2003; Xue-Zhong 2004) like patents are often considered, while in other cases what may count is the application of measures to prevent losses and leakages of previous knowledge possessed by the individuals operating in the company (Olander et al. 2011). Generally speaking, the problem of knowledge protection has become more and more critical because companies need to establish increasingly large networks with suppliers, business partners and clients, and to exchange a lot of precious knowledge with them, which raises the issue of if and how it is really possible to protect the value of the created knowledge (Norman 2001; Lee et al. 2007; Bolisani et al. 2013). Consequently, the strategic position of a company in relation to the protection of its knowledge assets affects and is influenced by the competitive positioning that it has or wants to have in the market. Also, there are cases where a company must reflect on the impossibility to protect knowledge effectively, and on the opportunities that are given by an opposite strategy which is called *open knowledge disclosure* (Pénin 2007): here, a company decides to anticipate possible leakages of precious elements of knowledge produced internally by deliberate exchanging it with others. The assumption is that we may have to new advantages in terms of network relationship with external partners and not only disadvantages coming from knowledge losses or misappropriation. Indeed, a balanced mix of protection and disclosure is one of the ingredients of the new strategies of *open innovation* (Chesbrough et al. 2006) that have started to appear since the last decade.

A fourth point is about a strategy to implement a *culture of knowledge sharing*. A company can decide different orientations: for example, a highly structured organization may be preferred, where employees mainly communicate along the hierarchical lines and use formal ways of storing and transferring knowledge; or conversely, a policy of knowledge sharing between individuals across the organization (Argote 1999; Dougherty 1999) may be decided, which allows flexible communications and avoids the creation of barriers and idiosyncratic "knowledge islands" (Franz et al. 2002).

In principle, we may say that the examples reported here may be just considered as part of a company strategy: indeed, a business plan of a company can have a section referring to one or some of the points mentioned above. The novelty of the notion of knowledge strategy is that these issues become central and are the object of a special analysis and consideration, and not simply a marginal or ancillary element of a more general business strategy.

To better explain the definition of knowledge strategy we can make a direct comparison with a company strategy in terms of detailed objects and contents (Fig. 5.1).

In short, as reported in Table 5.2, we may say that the typical contents of a company strategy can be detailed in terms of decisions about these main elements or dimensions, decisions which are deemed essential for achieving successful results and for fitting the vision that executives have of the company in the future competitive markets (Andrews 1971):

- the future *goals* of a company (i.e. the *vision* of the company's future)
- the *plans* that should be developed to achieve these goals
- and the *resources* that are needed to implement these plans.

The same can be said as regards a knowledge strategy, but with some essential differences. We may argue that a *knowledge strategy is about formulating a goal regarding knowledge and intellectual resources of a company* and, consequently, defining the way and actions to achieve these goals (Halawi et al. 2006). Therefore, what defines a knowledge strategy is the *goal* in terms of knowledge resources, the *plans* about how to achieve, manage and deliver these resources, and the *internal and external sources and structures* that the company will need.

Table 5.2 provides more details and examples that explain the comparison between the two notions. In the common practice, the goals of a company strategy are generally expressed in terms of typical *business objectives* (such as for example: markets, profits, competitive positioning, etc.); the plans regard typical business

Table 5.2 Company and knowledge strategy: a comparison

	Company strategy	Knowledge strategy
Goals	Expected profits, market share, competitive position, economic and organizational performances, etc.	What we want or need to know, what we want or need to learn; expected outcomes of knowledge acquisition, creation or sharing; domains and forms of knowledge that are important, etc.
Plans	Operations plans, Marketing plans, R&D projects, financial plans, etc.	Coding/decoding processes, storing/delivering methods, learning/training methods; sharing processes; etc.
Resources	Structures (offices, plants, etc.); infrastructures (e.g. IT systems); people (e.g. workforce, specialists, etc.); money	Internal/external sources of knowledge; IT systems (KM systems); KM people (knowledge officers, experts, etc.)

activities or core processes that are designed (i.e.: operations, marketing, R&D, etc.); and the resources consist of structures (plants, offices, physical assets), infrastructures (e.g. IT systems), people, and money.

By transposing these ideas to the case of knowledge strategy, we can argue that this can be expressed in terms of:

- *goals*, i.e.: *what we need or want to know or learn* in order to fit our future idea of the company, and what *we expect to achieve* for the benefit of our company by means of the processes of knowledge acquisition, generation, acquisition, and sharing
- *plans*, namely: the decisions about *practical implementation* of processes related to the management of knowledge (i.e. how to code/decode pieces of knowledge that, for example, need to be transformed into documents or routines; what methods of learning and training have to be employed for better achieving the goals of a knowledge strategy; domains and forms of knowledge that are considered strategic—for example: technical rather than market knowledge; tacit or explicit, etc.)
- *resources*, relating to decisions about structures and infrastructures that are deemed to be necessary for the success of a knowledge strategy, i.e.: what sources of knowledge are considered relevant and important to invest in; what IT systems (i.e. Knowledge Management Systems) are the most appropriate for the strategy; what knowledge management structures are needed (i.e.: offices to arrange, KM specialists to appoint, key users, etc.).

5.2 Knowledge Strategy or KM Strategy?

As we have recalled before, a knowledge strategy mainly focuses on the management of knowledge resources and intellectual capital. In other words, on the knowledge management capabilities of a firm. It is for this reason that, in the literature, another term is often used, which is *knowledge management strategy* or *KM strategy*. Since knowledge is also the *object* of KM practices, we may say that a KM strategy refers to the *guidelines, goals, resources, and long-term plans* of KM programs in a company. So, it may be wondered if there is really a need for another different definition of *KM strategy* and if there can be a clear distinction from knowledge strategy.

It is important to notice that, just like happened to knowledge strategy, on the notion of KM strategy scholars do not agree completely. As is well documented by Shannak et al. (2012), there are at least three different meanings associated with this term: KM strategy as *KM implementation strategy*, KM strategy as *approach to KM*, and KM strategy as *knowledge strategy*.

In the first meaning (KM strategy as KM implementation strategy), KM strategy is seen as the attempt to formulate intentional plans for explicitly managing knowledge (Sveiby 2001; Wenger 2004) and a sort of roadmap for the KM department of a company (Halawi et al. 2006). Accordingly, a KM strategy mainly deals with specific and detailed organizational, managerial and technical arrangements that a company adopts for its KM programs: in this view, a KM strategy would differ from knowledge strategy in terms of *level of generality* and of *practical application*: the latter deals with the way knowledge can support competitive advantage in general (Denford and Chan 2011), while the former would focus on specific implementation details of methods, managerial practices and infrastructures (deViron et al. 2014). A consequence of this view is that the formulation of a company's knowledge strategy should come prior to any KM plan.

In the second meaning (KM strategy as an approach to KM), the term is used by those who attempt to identify and categorize the possible general approaches to managing knowledge that a company can have. These approaches reflect the various perspectives that exist in the field (Donate and Canales 2012). For example, a popular classification distinguishes between codification and personalization KM strategy (Greiner et al. 2007; Hansen et al. 1999; Kumar and Ganesh 2011): *codification* (also named *system-oriented* KM strategy) focuses on capturing, codifying, storing and using explicit knowledge in a form that is compatible with a company's organizational objectives; *personalization* (or *human-oriented* KM strategy) has the goal to improve knowledge flows through networking and interactions. Another important classification distinguishes KM strategies on the basis of the sources used by a company to get key pieces of knowledge (Kim et al. 2003; Choi et al. 2008) that can be *external* (i.e. suppliers, customers, universities and public laboratories, services providers, etc.) or *internal* (namely, employees, R&D departments, etc.). In short, a company may be characterized by the kind of KM approach that it decides or intends to adopt: so, we can have companies that are said to adopt a personalization KM strategies, others a codification KM strategy, etc. Under this perspective, a KM strategy becomes a particular possible option for a knowledge strategy.

In the third meaning (KM strategy as knowledge strategy), the two notions are considered synonyms (Shannak et al. 2012). The implicit assumption is that companies that implement KM programs automatically adopt a knowledge strategy. Conversely, some authors argue that companies, aware of the importance of their knowledge base and the need to improve it (i.e., they know they require a knowledge strategy), implicitly adopt a KM strategy, but without declaring that explicitly (Garavelli et al. 2004).

It is difficult to choose what position is best—probably all them are important, because each underlines important issues of the process of strategic formulation of KM activities. In particular, it may be argued that a distinction between knowledge strategy and KM strategy remains useful, both in theoretical and practical terms, because it means that we may have two different steps of strategic formulation. A knowledge strategy refers to a company's general vision of its knowledge as an economic resource, its importance for the business and the competitive environment, the organizational attitude towards the sources of knowledge, and the methods to manage them. A knowledge strategy remains, therefore, at the same general level of the strategy (in practice, it can be even a part of it). Instead, a KM strategy recalls a direct and explicit decision of a company to adopt specific KM practices and systems. We may also say that the notion of KM strategy refers to a "lower" and more operative level, i.e. the planning and implementation of tools and operative methods for managing knowledge as economic resource: identification of key KM processes and assignment of related tasks to employees, selection of practices, and computer tools.

On the other hand, it is also clear that knowledge strategy and KM strategy are strictly connected to one another. Formulating a knowledge strategy—i.e., the overall vision of a company about its knowledge resources and their importance for competitiveness—is a necessary ingredient for implementing practical tools and methods to manage knowledge—i.e., the KM strategy. Using the synthetic formulation of Tiwana (2010, p. 103), we may conclude that "*knowledge drives strategy and strategy drives knowledge management.*". For this reason, in the rest of this book, we will accept this ambiguity of terms and the partial overlapping of the two definitions, but we will underline the differences when they are important.

5.3 Knowledge Strategy and Company Strategy: Which Comes First?

The existence of a close relationship between knowledge and strategy has been widely recognized by scholars well before the upsurge of interest in KM (Kogut and Zander 1992). As we have seen before, the idea of knowledge as a strategic weapon is not new (Kasten 2007), but it is the development of the knowledge-based view of the firm (Grant 1996; Sveiby 2001) and, especially, the growing interest in KM that have made knowledge strategy a key topic of analysis. As underlined by Snyman and Kruger (2004, p. 5), "knowledge (as a strategic resource) has an enabling role to play in the formulation of winning strategies. The true power of knowledge lies in its ability to positively influence, and enable, the business strategy. Synergy between the business strategy and the knowledge management strategy is thus essential."

The connection between knowledge strategy and company strategy can be seen in different terms. For example, in relation to the *business performances* or in terms of the *competitive positioning*.

Knowledge strategy and company strategy are clearly linked by the common goal to improve the performances of a company, goal whose achievement can be at risk if these two strategies don't fit (Greiner et al. 2007; Hofer-Alfeis 2003). Further developing the works of others, Greiner et al. (2007, p. 5) affirm that the management of knowledge "should provide a competitive advantage for the organization", and for this reason, any activity of knowledge management "should be tightly related to objectives and business strategies of the organization or subunit of the organization" because otherwise, failing "to add value to the organization, it is only cost intensive, useless, or even counterproductive". In short, a knowledge strategy, whose ultimate purpose is to provide guidance to the management of knowledge resources, can't be detached from a company strategy; consequently, the formulation of a knowledge strategy is a process that can't be disjointed from the formulation of the company strategy (Eisenhardt and Santos 2002).

Therefore, it can be said that the decisions of a company about its knowledge strategy can be connected with the organization's competitive positioning (Zack 1999a): on the one hand, "an organization's strategic context helps to identify knowledge management initiatives that support its purpose or mission, strengthen its competitive position, and create shareholder value. Intuitively, it makes sense that the firm that knows more about its customers, products, technologies, markets and their linkages should perform better"; on the other hand, while "strategic management models traditionally have defined the firm's strategy in terms of its product/market positioning—the products it makes and the markets it serves", a resource-based view and, later, a knowledge-based approach suggests that "firms should position themselves strategically based on their unique, valuable and inimitable resources and capabilities rather than the products and services derived from those capabilities.". In other words, while a company strategy generally implies a future vision of the company in terms of its product/market/technology positioning compared to competitors, a knowledge strategy refers to its relative position in terms of knowledge possessed, learnt, shared, and/or exploited (Zack 1999a) compared to its competitors.

If we accept that there is a strict relationship between knowledge strategy and company strategy, there is a second issue that is important to consider: what comes first, company strategy or knowledge strategy? As we will see, this question is important especially in terms of strategic planning, i.e. the process (and related issue) of formulation of a strategy and the models or approaches that can be used for this. More precisely, the question is: which of the following two is the case?

(a) Does a firm first define its company strategy, and only later (and consequently) its knowledge strategy, that therefore just becomes a component of the former to specify special goals or plans in terms of knowledge resources?
(b) Or vice versa: can the formulation of a knowledge strategy be placed on a sort of higher level compared to that of the company strategy, a level where the general vision of the company is defined and only consequently the specific and practical details of the business (i.e. the company strategy) will be decided?

To analyze this question, for now let's consider the situation of a company where strategic planning is a formal activity, i.e. its results are in the form of a document, a business plan, a set of guidelines for the future months or years, etc. (as we will discuss later in this book, this is not always the situation which we are interested to consider, but let's assume that this is the case just for the sake of clarity). This means that case (a) "company strategy comes before knowledge strategy" implies that a part of the business plan of a company will focus on knowledge resources and related management practices, deriving from the general company strategy. In other words, strategists will first decide goals and plans in terms of products, markets, competitors etc. and, as a consequence, knowledge and KM needs will derive from these goals. For example, a company may decide that it is strategically important to develop a new product (*company strategy*) and, on the basis of that, clarifies what elements of knowledge are necessary to do that and how to develop or acquire these elements (*knowledge strategy*).

Case (b) "knowledge strategy comes before company strategy" implies that, to plan a strategy, executive must firstly reflect on their future idea of the company in cognitive terms (i.e. what we want or need to know to be successful in the future), and this can later transform into new business opportunities in more traditional terms. For example, a company first decides that it is, for some reason, more convenient to acquire some knowledge from outside rather than to completely develop it internally (i.e., this is the view in terms of *knowledge strategy*), and this can later become a new strategy of open innovation for developing new products in collaboration with external partners (i.e. this turns into the *company strategy*).

Case (a) has long been the most popular in the literature: many important scholars that analyzed the strategic importance of knowledge tended to see a knowledge strategy as a component of the company strategy (a key component but, nonetheless, a component) (Grant 1996; Hansen et al. 1999; Zack 1999b; Earl 2001). Generally speaking, it was often assumed that when a company articulates its strategic business plans, the knowledge needed to fulfill this business strategy are analyzed and compared with the actual knowledge, then knowledge gaps are revealed and, finally, this may become a knowledge strategy and is incorporated as part of its "general" business plan (Zack 1999a).

The literature has started to consider case (b) more recently, but with increasing attention. Given the growing importance of KM practices in companies, many scholars have started to see KM as an "enabler to achieve strategic business objectives" (Halawi et al. 2006, p. 392). Therefore, the definition of a strategy for managing knowledge resources can also become new business opportunities and, therefore, strategic options in terms of the business. Practical management models—like for instance the Balanced Scorecard (Kaplan and Norton 2001), that represents a way to monitor the implementation of a company strategy—have also been adapted and enhanced to consider intellectual capital and, more generally, knowledge resources. For example, the "Skandia Navigator" model (Skandia 1996) was proposed as a new way to measure the strategic success of a company based on its intellectual capital, i.e. knowledge (Skyrme and Amidon 1998). Therefore, knowledge strategy and company strategy are placed at least on the same level.

Another particularly important point is the upsurge of interest on Knowledge-Intensive Business Services (KIBS). KIBS include companies whose importance is now recognized in the modern economy, such as e.g.: IT services, business consulting, financial or legal advisors, marketing and communications, media companies, etc. Particularly, these companies have been regarded as a distinctive trait of the knowledge economy and as such they have been the subject of an increasing number of studies (Doloreux et al. 2010). According to the extant literature, the functioning and the innovative capability of KIBS are strictly associated with the cognitive assets they possess (Miles et al. 1995; Strambach 2008; Muller and Doloreux 2009). First, their main production factor and outcome consists of knowledge, directly delivered to customers in the form of consulting, or embedded in artifacts and services. Second, their activity is mostly based on the exploitation of the specializations, skills and knowledge of their employees (i.e. knowledge workers, Alvesson 2004). Third, the provision of knowledge-intensive services requires an in-depth interaction between supplier and user, both involved in cognitive exchanges and learning processes (Bettencourt et al. 2002). Fourth, the delivery of such services involves a process of problem solving in which KIBS companies adapt their knowledge to the specific requirements of individual clients. Fifth, they often act as interfaces between global sources of knowledge and end users (Smedlund 2006). Finally, their innovative capability is directly connected to the acquisition, processing, capitalization and delivery of new knowledge (Amara et al. 2008).

The application of the notion of knowledge strategy to these companies overcomes the limited focus of "strategy to manage knowledge resources" that we have seen before: knowledge resources *are* the business. Formulating a knowledge strategy in KIBS (i.e. what we need to know, how we develop the capabilities of our knowledge workers, what knowledge we need to protect or exchange with others, etc.) can be seen as the planning of the company strategy, i.e. how R&D programs will be planned based on what we plan to learn in the future (Fratostiteanu 2010), how innovations will be produced (Forcadell and Guadamillas 2002), what new products and services will be delivered (Leiponen 2005), what modalities of interaction with customers will be implemented (Scarso and Bolisani 2010; Landry et al. 2012). Therefore, in this case, we can see company strategy as a derivation of knowledge strategy.

Clearly, the two cases analyzed here represent two extreme and opposite views, and in the real life, it is likely that companies combine both views. One case or another may prevail considering the characteristics of the company (i.e.: products, markets, structure, management style, etc.). Suppose that we can place companies in a continuum from the two extremes (Fig. 5.2, where companies are represented by means of triangles): in the left part of the picture we have companies where the management of tangible resources strongly prevails (essentially, traditional industrial manufacturers)—case a), and in the right part companies where the management of intangible or "knowledge-based" resources prevails (knowledge-intensive companies, roughly speaking KIBS) case b).

This different "composition" of a company's business may explain why companies of type (a) will tend to adopt a traditional view of company strategy, while for companies of type (b) the notion of knowledge strategy tends to be more important. Real companies can be placed in between these two cases, which helps to understand the relative importance and priority of company or knowledge strategy for their particular case. This aspect will also be analyzed in the next chapters.

Another perspective that can be used to investigate the circular relationship between company strategy and knowledge strategy is depicted in Fig. 5.3. The

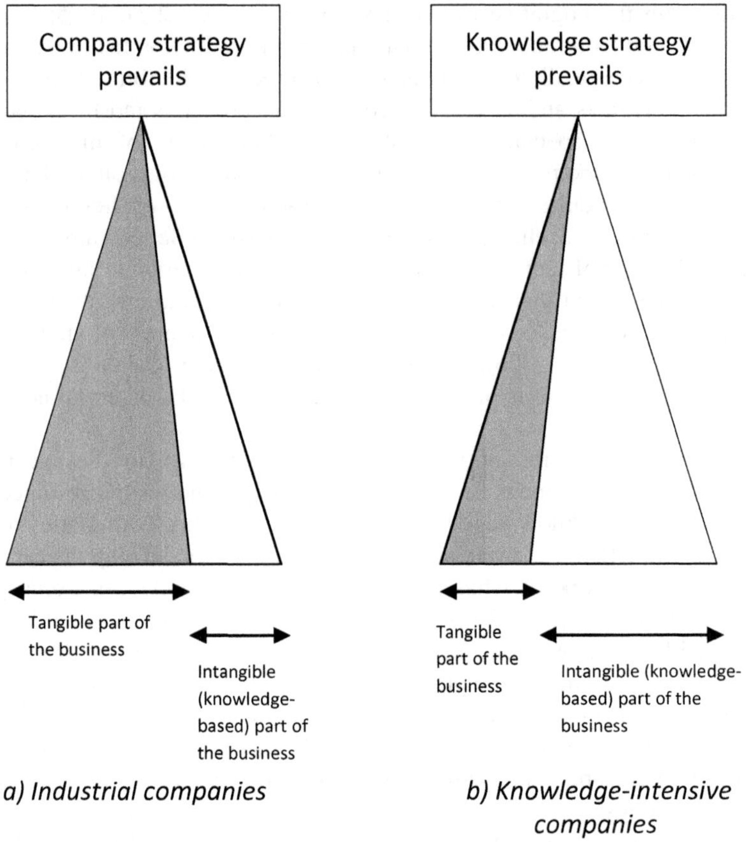

Fig. 5.2 Prevalence of company strategy or knowledge strategy in different kinds of companies

bottom part of the picture describes that knowledge has a potential for value creation in a company: both internally (for example, by means of KM processes, a company can try to make its processes more efficient and effective) and externally (for instance, a company may sell "knowledge-intensive" services to its clients, or simply incorporates highly valuable knowledge into its high-tech products: in other words, the knowledge produced internally becomes value for the market). Focusing on this leads to consider that formulating a business strategy also means to include analysis and decision on how knowledge can be transformed into value for the company. The implicit assumption here is that *business strategy comes prior to knowledge strategy.*

The upper part of the picture describes the issue that was recalled in the previous chapters: the possibility itself of adopting a company strategy is challenged by various factors related to the knowledge that we can have of the company and the external word. Unpredictability of the environment makes it difficult to formulate sensible forecasts about the environment (i.e. markets, competitors, innovations,

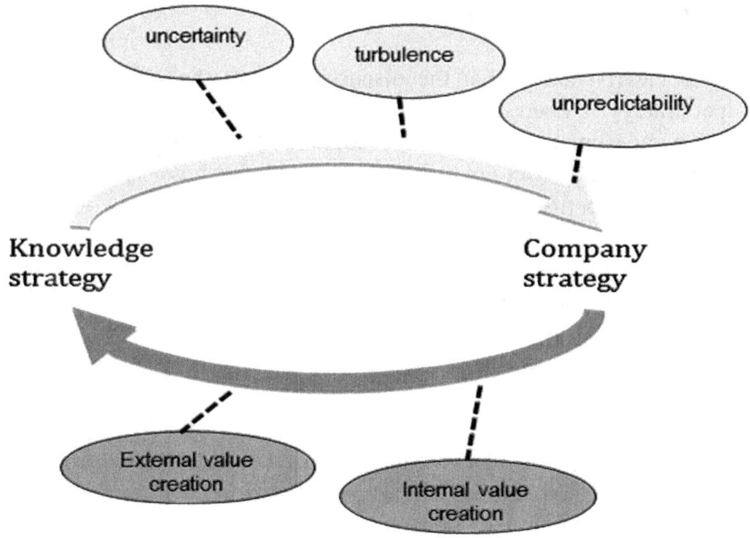

Fig. 5.3 Circular relationship between knowledge strategy and company strategy

etc.) and, especially, to decide goals that can be reasonably achieved. Turbulence (i.e. continuous changes in the environment or in the internal components of the organization) causes instability and, therefore, possible failures in the plans that a strategy includes. Uncertainty means that it is difficult to collect and analyze the information that is needed to formulate a company strategy (that, by definition, is about the future which is never certain), not only about markets and competitors but also regarding the whole organization itself.

Therefore, deciding goals, plans and resources of a company strategy is made difficult by the intrinsic instability of systems and phenomena that need to be under consideration. In this case, an appropriate knowledge strategy can become a key prerequisite of the company strategy: for example, analyzing and deciding what the company already knows, what it doesn't know (yet), what it should know and how this knowledge can be acquired, can help strategists to implement a capability to learn fast and to face unpredicted challenges more easily. In other words, this means deciding an appropriate strategy to transform a company into a *learning organization*, capable to quickly react to the challenges uncertainty and turbulence. According to some scholars, this appears to be one of the major reasons of *prevalence of a knowledge strategy* over a business strategy (Zack 2003; King 2001).

5.4 A Tentative Summary

Our analysis clearly shows that the discussion about the possible definition of knowledge strategy is open, and the notion is still fluid and debatable. However, what we have underlined in this chapter is that the idea of a knowledge strategy can be useful, and especially (but not only) when knowledge has a special place among the assets and competitive resources of a company. This may explain why a "separate" strategic formulation becomes vital: in other words, it becomes important to reflect on the strategy for managing knowledge as a sort of "distinct exercise" that a company may do in comparison to the classic idea of company strategy. Indeed, treating knowledge strategies explicitly implies to adopt new conceptual categories, classifications and methods compared to the traditional elements of company strategies.

Particularly, we have seen that the increasing use of KM practices and the adoption of KM programs in companies necessarily imply a reflection on the strategic meaning of KM. So, KM is not simply a low level operational activity that implies short term or tactical decisions, but tends to become an essential ingredient of a company's business, because it can influence organizational performances and innovation capabilities. For this reason, KM requires special attention in strategic terms, and can even influence the overall strategic orientation of a company. As we will see in the following chapters, all this leads to reflect on specific frameworks and models to understand the place of knowledge strategies in business, and on special approaches to guiding strategic planning.

In short, knowledge strategy is now strictly connected to the more traditional notion of strategy in business. We have seen that the two concepts focus on different elements of a company's business—although for some companies (for example, for KIBS), a company strategy substantially *is* a knowledge strategy. As we have discussed here, it is not easy to clarify if companies can consider a knowledge strategy simply as a component of its company strategy, or a separate and primary element of strategic formulation. We have argued that the presence of uncertainty, turbulence and unpredictability of the environment and the organization itself, challenges the possibility to formulate a company strategy and, in this case, a reflection on how a company can be put in a condition to learn fast and react quickly to unforeseen changes may become even more important. In this case, it is an appropriate knowledge strategy (i.e. deciding how a company can improve its learning capability) that can be the vital, essential because it is a way to transform companies into learning organizations.

The discussion that we have proposed in this chapter can be important and interesting for researchers, but is not purely theoretical: if a knowledge strategy finds a special place in strategic thinking, this implies that we need to adopt proper planning methods and to reflect on new strategic options that go beyond the traditional classifications that we are used to employing in strategic management. This is what we are going to treat in the next chapter.

References

Alvesson, M. (2004). *Knowledge work and knowledge-intensive firms*. Oxford: Oxford University Press.

Amara, N., Landry, R., & Traorè, N. (2008). Managing the protection of innovations in knowledge-intensive business services. *Research Policy, 37*, 1530–1547.

Andrews, K. R. (1971). *The concept of corporate strategy*. Homewood, IL: Dow Jones-Irwin.

Argote, L. (1999). *Organizational learning, retaining and transferring knowledge*. Boston: Kluwer Academic.

Bettencourt, L. A., Ostrom, A. L., Brown, S. W., & Roundtree, R. I. (2002). Client co-production in knowledge-intensive business services. *California Management Review, 44*(4), 100–128.

Bierly, P., & Chakrabarti, A. (1996). Generic knowledge strategies in the US pharmaceutical industry. *Strategic Management Journal, 17*(Winter), 123–135.

Bolisani, E., Paiola, M., & Scarso, E. (2013). Knowledge protection in knowledge-intensive business services. *Journal of Intellectual Capital, 14*(2), 192–211.

Chesbrough, H., Vanhaverbeke, W., & West, J. (2006). *Open innovation: Researching a new paradigm*. Oxford: Oxford University Press.

Choi, B., Poon, S. K., & Davis, J. G. (2008). Effects of knowledge management strategy on organizational performance: A complementarity theory-based approach. *Omega, 36*(2), 235–251.

Cillo, P. (2005). Fostering market knowledge use in innovation: The role of internal brokers. *European Management Journal, 23*(4), 404–412.

Coakes, E., Amar, A. D., & Granados, L. M. (2010). Knowledge management, strategy, and technology: A global snapshot. *Journal of Enterprise Information Management, 23*(3), 282–304.

Davenport, T. H., & Harris, J. G. (2007). *Competing on analytics: The new science of winning*. Boston: Harvard Business School.

De Luca, M. D., & Atuahene-Gima, K. (2007). Market knowledge dimensions and cross-functional collaboration: Examining the different routes to product innovation performance. *Journal of Marketing, 71*(1), 95–112.

Denford, J. S., & Chan, Y. E. (2011). Knowledge strategy typologies: Defining dimensions and relationships. *Knowledge Management Research & Practice, 9*(2), 102–119.

deViron, F., De Jaegere, T., Lederer, T., & Vas, A. (2014). Exploring knowledge strategy within a knowledge-intensive organisation: A case study approach. *International Journal of Information Technology and Management, 13*(4), 264–284.

Doloreux, D., Freel, M. S., & Shearmur, R. G. (Eds.). (2010). *Knowledge-intensive business services: Geography and innovation*. Aldershot: Ashgate Publishing.

Donate, M. J., & Canales, J. I. (2012). A new approach to the concept of knowledge strategy. *Journal of Knowledge Management, 16*(1), 22–44.

Dougherty, V. (1999). Knowledge is about people, not databases. *Industrial and Commercial Training, 31*(7), 262–266.

Earl, M. (2001). Knowledge management strategies: Toward a taxonomy. *Journal of Management Information Systems, 18*(1), 215–233.

Eisenhardt, K. M., & Santos, F. M. (2002). Knowledge-based view: A new theory of strategy. *Handbook of Strategy and Management, 1*, 139–164.

Forcadell, F. J., & Guadamillas, F. (2002). A case study on the implementation of a knowledge management strategy oriented to innovation. *Knowledge and Process Management, 9*(3), 162–171.

Franz, M., Freudenthaler, K., Kameny, M., & Schoen, S. (2002). The development of the Siemens knowledge community support. In T. H. Davenport & G. J. B. Probst (Eds.), *Knowledge management case book* (pp. 147–159). Berlin: Wiley.

Fratostiteanu, C. (2010). Guidelines for promoting science, technology and technical–scientific creativity, by analyzing the companies' performances, in the context of the globalized economy. *Journal of Applied Economic Sciences (JAES), 13*, 247–257.
Garavelli, C., Gorgoglione, M., & Scozzi, B. (2004). Knowledge management strategy and organization: A perspective of analysis. *Knowledge and Process Management, 11*(4), 273–282.
Grant, R. M. (1996). Toward a knowledge-based theory of the firm. *Strategic Management Journal, 17*(S2), 109–122.
Greiner, M. E., Böhmann, T., & Krcmar, H. (2007). A strategy for knowledge management. *Journal of Knowledge Management, 11*(6), 3–15.
Halawi, L. A., McCarthy, R. V., & Aronson, J. E. (2006). Knowledge management and the competitive strategy of the firm. *The Learning Organization, 13*(4), 384–397.
Hansen, M. T., Nohria, N., & Tierney, T. (1999). What's your strategy for managing knowledge? *Harvard Business Review, 77*, 106–116.
Hofer-Alfeis, J. (2003). Effective integration of knowledge management into the business starts with a top-down knowledge strategy. *Journal of Universal Computer Science, 9*(7), 719–728.
Holsapple, C. W., & Jones, K. G. (2007). Knowledge chain activity classes: Impacts on competitiveness and the importance of technology support. *International Journal of Knowledge Management (IJKM), 3*(3), 26–45.
Johannessen, J. A., Olsen, B., & Olaisen, J. (1999). Aspects of innovation theory based on knowledge-management. *International Journal of Information Management, 19*(2), 121–139.
Kaplan, R. S., & Norton, D. P. (2001). The strategy-focused organization. *Strategy and Leadership, 29*(3), 41–42.
Kasten, J. (2007). *Knowledge strategy and its influence on knowledge organization.* In Proceedings of the North American Symposium on Knowledge Organization, Vol. 1 (online). http://dlist.sir.arizona.edu/1907
Kim, Y. G., Yu, S. H., & Lee, J. H. (2003). Knowledge strategy planning: Methodology and case. *Expert Systems with Applications, 24*(3), 295–307.
King, W. R. (2001). Strategies for creating a learning organization. *Information Systems Management, 18*(1), 12–20.
Kogut, B., & Zander, U. (1992). Knowledge of the firm, combinative capabilities, and the replication of technology. *Organization Science, 3*(3), 383–397.
Kraus, S. (2007). Strategic planning in new ventures and young SMEs. In C. Wankel (Ed.), *21st century management: A reference handbook.* Thousand Oaks: Sage Publications.
Kumar, A. J., & Ganesh, L. S. (2011). Balancing knowledge strategy: Codification and personalization during product development. *Journal of Knowledge Management, 15*(1), 118–135.
Landry, R., Amara, N., & Doloreux, D. (2012). Knowledge-exchange strategies between KIBS firms and their clients. *The Service Industries Journal, 32*(2), 291–320.
Lee, S. C., Chang, S. N., Liu, C. Y., & Yang, J. (2007). The effect of knowledge protection, knowledge ambiguity, and relational capital on alliance performance. *Knowledge and Process Management, 14*(1), 58–69.
Leiponen, A. (2005). Organization of knowledge and innovation: The case of Finnish business services. *Industry & Innovation, 12*(2), 185–203.
López-Nicolás, C., & Meroño-Cerdán, Á. L. (2011). Strategic knowledge management, innovation and performance. *International Journal of Information Management, 31*(6), 502–509.
McManis, C. R. (2003). Intellectual property, genetic resources and traditional knowledge protection: Thinking globally, acting locally. *Cardozo Journal of International and Comparative Law, 11*(2), 547–583.
Miles, I. (2005). Knowledge intensive business services: Prospects and policies. *Foresight, 7*(6), 39–63.
Miles, I., Kastrinos, N., Bilderbeek, R., & den Hertog, P. (1995). *Knowledge-intensive business services: Users, carriers and sources of innovation* (EIMS Publication No. 15). Luxembourg : European Innovation Monitoring System (EIMS).

Mládková, L. (2014). Knowledge strategy: Key player or relict of the past? *Procedia-Social and Behavioral Sciences, 150*, 628–636.

Muller, E., & Doloreux, D. (2009). What we should know about knowledge-intensive business services. *Technology in Society, 31*, 64–72.

Norman, P. M. (2001). Are your secrets safe? Knowledge protection in strategic alliances. *Business Horizons, 44*(6), 51–60.

Olander, H., Hurmelinna-Laukkanen, P., & Heilmann, P. (2011). Do SMEs benefit from HRM-related knowledge protection in innovation management? *International Journal of Innovation Management, 15*(03), 593–616.

Pénin, J. (2007). Open knowledge disclosure: An overview of the evidence and economic motivations. *Journal of Economic Surveys, 21*(2), 326–347.

Scarso, E., & Bolisani, E. (2010). Knowledge-based strategies for knowledge intensive business services: A multiple case-study of computer service companies. *Electronic Journal of Knowledge Management, 8*(1), 151–160.

Schulz, M., & Jobe, L. A. (2001). Codification and tacitness as knowledge management strategies: An empirical exploration. *The Journal of High Technology Management Research, 12*(1), 139–165.

Senker, J. (1995). Tacit knowledge and models of innovation. *Industrial and Corporate Change, 4* (2), 425–447.

Shannak, R. O., Ra'ed, M., & Ali, M. (2012). Knowledge management strategy building: Literature review. *European Scientific Journal, 8*(15), 143–168.

Skandia. (1996). *Supplement to Skandia's 1995 Annual Report*. Stokholm: Skandia.

Skyrme, D. J., & Amidon, D. M. (1998). New measures of success. *Journal of Business Strategy, 19*(1), 20–24.

Smedlund, A. (2006). The roles of intermediaries in a regional knowledge systems. *Journal of Intellectual Capital, 7*(2), 204–220.

Snyman, R., & Kruger, C. J. (2004). The interdependency between strategic management and strategic knowledge management. *Journal of Knowledge Management, 8*(1), 5–19.

Spender, J. C., & Grant, R. M. (1996). Knowledge and the firm: Overview. *Strategic Management Journal, 17*(S2), 5–9.

Strambach, S. (2008). Knowledge-Intensive Business Services (KIBS) as drivers of multilevel knowledge dynamics. *International Journal of Services Technology and Management, 10*(2/3/4), 152–174.

Sveiby, K. E. (2001). A knowledge-based theory of the firm to guide strategy formulation. *Journal of Intellectual Capital, 2*(4), 244–258.

Tiwana, A. (2010). Systems development ambidexterity: Explaining the complementary and substitutive roles of formal and informal controls. *Journal of Management Information Systems, 27*(2), 87–126.

von Krogh, G., Nonaka, I., & Aben, M. (2001). Making the most of your company's knowledge: A strategic framework. *Long Range Planning, 34*(4), 421–439.

Walter, C., & Ribière, V. (2013). A citation and co-citation analysis of 10 years of KM theory and practices. *Knowledge Management Research & Practice, 11*(3), 221–229.

Wenger, E. (2004). Knowledge management as a doughnut: Shaping your knowledge strategy through communities of practice. *Ivey Business Journal, 68*(3), 1–8.

Xue-Zhong, Zhu. (2004). Study on the legal protection of traditional knowledge. *Journal of Central China Normal University (Humanities and Social Sciences) 3*(4).

Zack, M. (1999a). Developing a knowledge strategy. *California Management Review, 41*(3), 125–145.

Zack, M. H. (1999b). Managing codified knowledge. *Sloan Management Review, 40*(4), 45–58.

Zack, M. H. (2003). Rethinking the knowledge-based organization. *MIT Sloan Management Review, 44*(4), 67–71.

Zhou, K. Z., & Li, C. B. (2012). How knowledge affects radical innovation: Knowledge base, market knowledge acquisition, and internal knowledge sharing. *Strategic Management Journal, 33*(9), 1090–1102.

Knowledge and Strategy Formulation in a Turbulent World

6

In this chapter, we focus on strategy formulation, i.e. the process of formulation of a strategy for a company. For strategists, an ideal world is that where this process can take place systematically and with no ambiguity or uncertainties: goals are decided, data are collected and analyzed, and final decisions are taken and, later, implemented. Therefore, the intrinsic presence of sources of uncertainty and turbulence affect the way strategy formulation can be approached consistently.

In addition, the introduction of the notion of knowledge strategy poses new challenges. Can the notion of knowledge strategy be some help for strategists, or does it add new complications to strategy formulation? How can a knowledge strategy be formulated? Is knowledge strategy formulation somewhat special? And how is it related to that of a company strategy? These are the main questions we address in this chapter.

6.1 Strategists Under the Sword of Damocles

The legendary anecdote of the *Sword of Damocles* was first cited by the historian *Timaeus* of Tauromenium (today's Taormina, in Sicily) and later recalled by the Roman orator and writer Cicero in his philosophical writing *Tusculanae Disputationes*. In the classic European culture, the legend has become a classic metaphor of the ever-present risks faced by those people that are in a position of power; more generally, it denotes the sense of precariousness that necessarily affects the course of human events and decisions.

According to the story, Damocles, a noble courtier of Dionysius I (the tyrant of Syracuse, in the fourth century BC), never missed an opportunity to remind his king how fortunate he was as a great man of power, surrounded by luxury and magnificence. One day Dionysius, probably fed up with that harassment, offered to switch places with Damocles so that he could taste by himself what he insisted was a real fortune. Damocles happily accepted, and sat down in the tyrant's throne during a lavish banquet. However, Dionysius had arranged that a sharp sword hanged above

E. Bolisani, C. Bratianu, *Emergent Knowledge Strategies*, Knowledge Management and Organizational Learning 4, DOI 10.1007/978-3-319-60657-6_6

the throne, over Damocles' neck, and held just by a single hair of horsetail. When Damocles became aware of his risky situation, he begged Dionysius that he be allowed to leave the throne and go back to his own seat, because he no longer wanted to be so fortunate as his king, realizing the dangers that his great fortune and power also brought about.

This metaphor can be easily applied to the situation that we want to describe here: that of decision-makers and strategists in companies. We may say that they have great power, and what they decide can even influence the fate of the people that are subjected to their decisions. As is the tradition especially in strongly hierarchical organizations, when the top management formulates "the strategy", the rest of the company will follow, and all employees will need to implement the strategy and align to it as best as possible.

But at the same time, strategists must bear all the uncertainties and risks that are associated to their role. What will happen if their decisions are wrong? What if, due to their mistakes, the profits and even the survival of their company are at risk? Maybe they won't be asked to pay the price (indeed, we have many recent examples of failures of companies and banks that didn't lead to the misfortune of their CEOs), but in any case, taking risk-free decisions is clearly better that bearing risks of being remembered as the ones that made dreadful mistakes.

In this chapter, we focus on strategy planning, i.e. the process of formulation of a strategy for a company. Generally speaking, strategy planning or, as it is often called, strategic planning can be defined as the process through which a company strategy is decided and applied by declaring its goals, the modality to achieve them, and the other implementation details. For strategists, an ideal world is that where this process can take place systematically and with no ambiguity or uncertainties: goals are decided, data are collected and analyzed, final decisions are taken and, later, implemented. Therefore, the intrinsic presence of sources of uncertainty and turbulence affect the capability to provide sensible forecasts about the future (see Chap. 4) and, consequently, also reflects on the way strategic planning can be approached consistently.

The questions that we address in this chapter are substantially as follows: is there a planning approach that can help companies to face the uncertainties of a turbulent and unpredictable world? How can strategists save their role but, at the same time, take sensible decisions and formulate appropriate strategies for their company? And, especially important here: is the notion of knowledge strategy of some help for strategists? And how can a knowledge strategy be planned?

The literature has long emphasized how the awareness of uncertainty and environmental turbulence have progressively changed the way strategic planning is approached in companies. In this first section, we will provide an outline of the main problems of a rational approach to strategic planning. Also, we will discuss an opposite view of strategy, which refers to the notion of emergent strategy planning.

6.1.1 How to Formulate a Strategy? From Strategic Planning to Strategic Management, and Beyond

At the beginning of this chapter, a terminological clarification is necessary. Indeed, by analyzing the managerial literature, we would find several terms that refer to the formulation of a strategy in business (Hussey 2007). Strategic planning was one of the first terms used since the 1960s to indicate the activity aimed to formulate a long-term strategic plan for companies: generally speaking, it is the systematic and analytical process of defining a company strategy. Strategic management is a development of the previous concept, with some important differences: first, while strategic planning (Gluck et al. 1980) is more focalized on the external environment (i.e. a strategy is planned based on an analysis of markets, competitors, regulatory systems, etc.), strategic management also includes the consideration of internal elements of the organization; second, strategic management emphasizes not only the "hard" elements that are a necessary input of a process of analysis and planning (primarily markets, products, materials, and competitors), but also on "soft" elements (like e.g. culture, social environment, management style, etc.) that influence the formulation of a strategy. Later, in the notion of strategic management, others also see the consideration of personal creativity of strategists, that must develop a *vision* of the company, whose implementation is the ultimate purpose of a strategy: "Successful business strategies result not from rigorous analysis but from a particular state of mind. In what I call the mind of strategists, insight and a consequent drive for achievement, often amounting to a sense of mission, fuel a thought process which is basically creative and intuitive rather than rational" (Ohmae 1983, p. 4). More recently, another puzzling term is strategic thinking. Initially its use was inspired by the criticism to the term strategic planning expressed by authors such as Mintzberg (1994), term which, however, is used with different shades of meaning. Heracleous (1998, p. 31) summarizes the idea of strategic thinking arguing that "The purpose of strategic thinking is to discover novel, imaginative strategies that can re-write the rules of the competitive game; and to envision potential futures significantly different from the present". In this view, strategic planning is based on the analysis of a state and on sensible forecasts about the future, and is substantially a *reactive* process to the changing conditions of the environment; strategic thinking is an intuitive vision of a substantially unpredictable future, and a *proactive* process that aims to anticipate change.

Indeed, this "terminological instability" reflects the richness of the debate on strategies. In addition, at least for the goals of this chapter, it is difficult to prefer one or the other of these terms. Therefore, here we will use a "neutral" and generic term, and speak of "strategy formulation". We will however use specific words when appropriate.

There are many schools of thought about strategy formulation. By developing a previous taxonomy proposed by Mintzberg (1990), Elfring and Volberda (2001) classify these schools into some main streams.

The *prescriptive* schools include authors such as Andrews, Ansoff and Porter, and consider the formulation of a strategy as an effort to "design" it, to "plan" it, or to decide where to "position" a company in the competitive environment. All these schools have some common traits: they see strategy formulation as a *top-down* process (i.e.: decisions are taken by the top management, and implemented by the ground operatives), based on a well-organized, systematic and analytical planning activity, and under the assumption of *perfect rationality* (or at least, perfect-enough rationality). The environment is seen as substantially constant, and analyzable rationally. The prescriptive schools substantially adopt a conceptual model of a company as a system that can and should be controlled, and strategy formulation as an effort of rationalization based on appropriate knowledge of this system.

The *descriptive* schools refer to contributions that include several different authors, from Schumpeter to March, from Simon to Normann, and also others. Rather than proposing a "best way" to formulate strategies, these schools observe the way real companies formulate their strategy and, based on these empirical observations, to analyze the factors that can influence how a strategy is actually decided. These schools are, indeed, quite different from one another but, according to Mintzberg (1990), they have some common traits: for example, they underline the intrinsic instability of the environment, which necessarily affects strategy formulation. Under a condition of bounded rationality and uncertainty, a company strategy is seen as a *visionary* effort of entrepreneurs or managers rather than a rational process; as based on *signals* that spontaneously come from the ground operations (bottom-up) rather than on *a pre-build model* of the company; as a result of *balances of power and coalitions* in the company; and as depending on *shared values and ideologies* of individuals which also form the *collective culture* of a company. Mintzberg didn't include himself in this group of authors, but we may say that his criticism against a prescriptive view of strategy formulation and his notion of emergent strategy (Mintzberg and Waters 1985) has much in common with these schools. We will recall this later in this chapter.

In addition to these two important groups, Elfring and Volberda (2001) add another group of "emerging" schools (at that time) that include: the *boundary school* (with authors such as Mahoney, Jarillo and Powell) that see the focus of strategies as the development of internal core competences and, when this is not possible or easy, the networking with external partners; the *dynamic capability school* (that refers to seminal studies by Schoemaker, Barney, Teece, and others) that sees strategic management as a collective effort of learning; and the *configurational school* (that includes the works of Mintzberg and colleagues) that considers strategic formulation as an "episodic" process depending on the specific organizational environment characterizing each a company.

To this classification, a group of more recent schools can be added. These schools emphasize some points that are particularly important in today's economy. The *strategy-as-practice* school, based on seminal works of Whittington (1996), emphasizes the idea of *strategizing* rather that a specific strategy as an ultimate output, and sees it as a *social and collective process* that involves learning at all levels of the organization and where executives are more *mediators* of the different contributions rather than old-style decision-makers. The *strategic agility* school (Roth 1996) emphasizes the necessity that companies become able to adapt quickly to a turbulent and fast-changing environment. For this, executives must adopt a flexible management style, that exploits the capability to leverage on the distributed "intelligence" of the organization, and on the capability to learn by all employees.

In practical terms, all these different viewpoints lead to different suggestions or guidelines for managers. In addition, it reveals that the process of strategy formulation is intrinsically complex and can imply a mix of different approaches on the basis of contingent situations. To analyze the possible approaches to strategy formulation, we will simply refer to two extreme ideal types, one that we denote *deliberate or rational approach*, and the other that we call *emergent*.

6.1.2 The Appeal of Deliberate and Rational Strategy Formulation

The adjective "deliberate" implies that an explicit and systematic process, aiming at formulating "the official strategy" of the company, takes place periodically. "Rational" means that strategy formulation is based on structured schemes and procedures that guide the decision-making process.

A deliberate and rational approach implies that companies can and should set their long-term goals rationally and deliberately: based on that, resources and plans to pursue these goals are defined. Goals, plans, and resources originate from precise intentions of the company and good knowledge of internal and external environment, which are seen to be predictable (at least at a sufficient degree).

All is decided by central leadership, progressively articulated in more detailed tasks that involve different parts of the organization, and later transmitted, in a top-down logic, to those that, in the company, must implement the strategy with operative actions (i.e. production, sales, deliveries, etc.). Essential pre-conditions are that the internal and external environments can be benign, controllable, and foreseeable, and there is full understanding, adhesion, and acceptance, by the different parts of the organization, to the tasks and processes that goals and plans require.

The idea of a deliberate and rational approach to strategy formulation has its origins in the concept of strategic planning as was originally introduced by authors, such as Chandler (1962), Ansoff (1965) and Drucker (1954). For Chandler (1962), the formulation of a strategy is one of the primary acts of executives, and influences the choices about the organizational structure of the company itself. Therefore, the activity of strategy formulation is so important that executives must be put in the best conditions to perform this duty in the most appropriate administrative environment, and they must have a *rational method* that helps them in this complex process. According to Drucker (1954), the formulation of a strategy is about taking decisions that anticipate the future risks in the long term, and this "requires that the work to be done to produce the desired future be clearly defined and clearly assigned". His definition is that formulating a strategy implies "the continuous process of making present risk-taking decisions systematically with the greatest

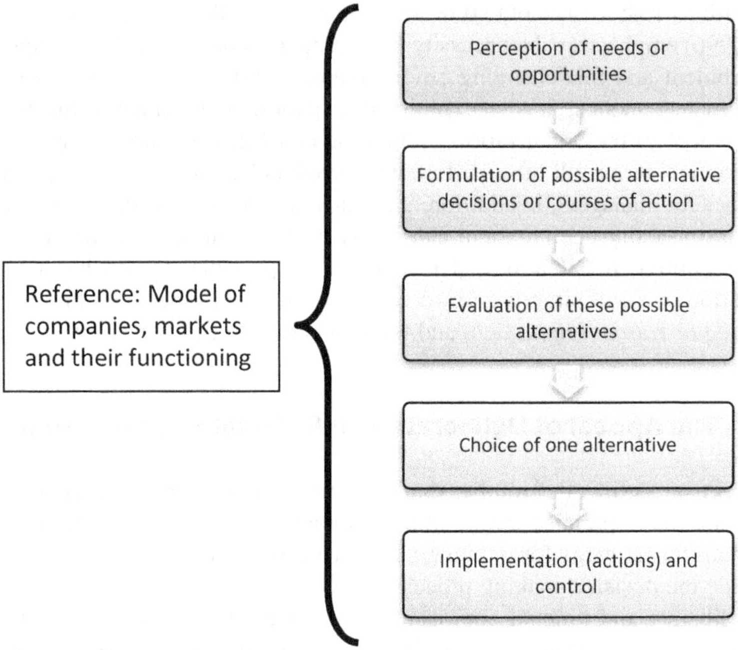

Fig. 6.1 The basic features of a deliberate and rational approach to strategy formulation

knowledge of their futurity; organizing systematically the efforts needed to carry out these decisions; and measuring the results of these decisions against the expectations through organized, systematic feedback". To summarize, strategist should be aware that the future is unpredictable, but they must face the challenge to take risky decisions with the purpose to lead the company towards the desired future and to minimize the risk of failure by systematic analysis of the achieved results.

Ansoff (1965) is one of the first authoritative authors to develop a structured method for strategy formulation. According to Martinet (2010) "Ansoff was deeply convinced that deliberate strategies are necessary to achieve long-term performances and targets". He used previously existing representations a rational decision-making process (see e.g. Simon 1960) consisting of a sequence of typical steps (Fig. 6.1) whose starting point is a clear reference model of how companies and markets function.

The idea of a deliberate and rational model was ideated in a time (just after the WWII) when world economy was growing fast and could be considered sufficiently "stable" and "predictable", especially for the big multinational companies that kept a strict control over the production cycle and the markets. Ansoff himself recognized that the full feasibility of a structured approach like that may be limited to these specific conditions, and may be too rigid in case of incumbent uncertainties, and proposed an "adaptive" approach consisting of an iteration of strategic decisions, actions, evaluations, and again decisions; indeed, he introduced elements of flexibility and feedback analysis in the complete formulations of his planning models (Ansoff 1965).

But in any case, today the idea itself of a deliberate and rational approach to strategy formulation remains quite popular in the managerial literature. Even the recent textbooks for managers (just to provide some examples, see Simerson 2011; Karami 2012; Ulwick 1999) are plenty of rational schemes and models that are expected to help executives in the analysis of markets and competitive forces, and in the formulation of company strategies.

Today, there is clear awareness that things can be complex and that uncertainties and fast-changing environments can seriously affect the effectiveness of predictions and of rationally-based decisions, that organizations are complex systems where many people can influence or be involved in decision making, and that there is the need to learn from experience and to change established viewpoints on societies, technologies or economies. But nonetheless, the appeal of a deliberate and rational approach to strategy formulation remains strong because its foundations provide a recognized landmark.

6.1.3 Emergent Strategy Formulation

After the pioneering studies of strategic planning, the dynamics of markets and technologies have soon challenged the possibility to formulate strategies rationally. Some authors have proposed to consider the effects of uncertainty and dynamic changes on managerial decisions: companies never have all the information required to set plans in advance, and new scenarios may emerge all the time. Also, those who criticize rational planning underline that, while this school "assumes that a correct strategy can only come about by means of frequent and systematic forecasting, planning and control ... in turbulent environment planning is, however, often insufficient and leads to rigidity. The annual planning rituals within an organization restrict its innovative potential; options are fixed and new options are not noticed" (Elfring and Volberda 2001, p. 4). What emerges is that strategists, sitting in their war room, may not have enough knowledge to predict what needs to be predicted, and/or to formulate and analyze all the possible strategic options for the entire company: so, it may be wiser to let the organization free to experiment new initiatives that can come from any part of the company.

A totally opposing view to that of rational planning is the notion of emergent strategy, whose earlier definition and development can be found in Mintzberg and Waters (1985) and Mintzberg (1987). A purely *emergent approach to strategy formulation* can be defined as one where actions result overtime, but in absence of deliberate intentions, clear leadership, and anticipated definition of goals. So, plans of a company result from an ex-post formalization and co-ordination of actions, decisions, and tasks that prove to be effective and beneficial to the organization. In substance, emergent decision making is based on a reactive organizational behavior and adaptive organizational learning (Senge et al. 1999). Opposing

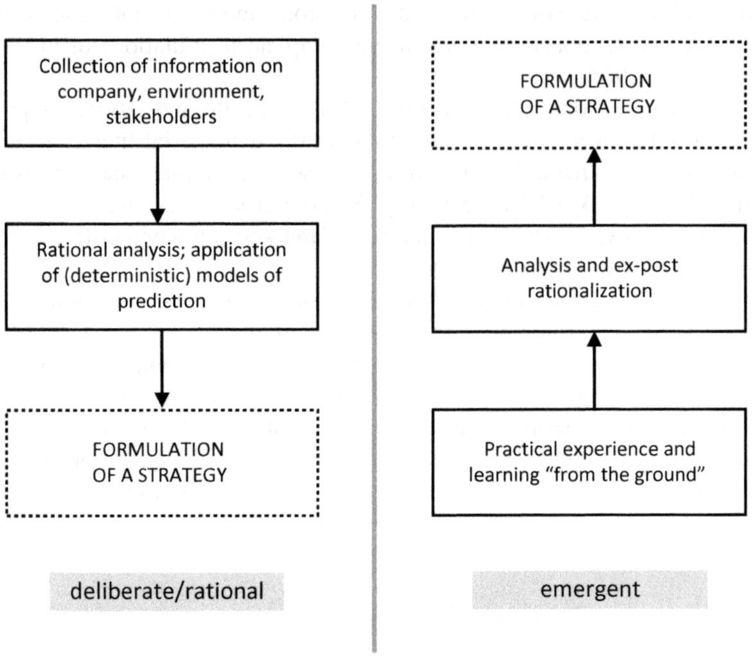

Fig. 6.2 Deliberate/rational vs. emergent strategy formulation

to the idea of a "grand" strategy, formulated by executives and next applied to the entire enterprise, the notion of emergent strategy recalls that strategies may be not always explicitly formulated, or at least, can come spontaneously, without aprioristic intentions, and may be approved afterwards by the management in a bottom-up approach (a "retrospective sense making"—cfr. Burgelman 1983; Bourgeois and Brodwin 1984). While the *deliberate strategies* can be better understood with the time metaphor of the moving observer (see Chap. 4 for details), *emergent strategies* resonate with the opposing metaphor of the flying time. That means that emergent strategies are reactions to the changeable business environment when the future comes to us in a different configuration than we have expected.

There is a radical change of perspective between the two mentioned approaches (Fig. 6.2 and Table 6.1). As we have mentioned, deliberate/rational strategy formulation assumes a structured process that takes places periodically and is based on clear and rational models. It is a top-down approach, with a clear distinction of competences (the top management formulates the strategy, and the operatives implement it) and precise time sequence (fist strategy formulation, later implementation). It assumes a "good-enough" knowledge of the environment and the company functioning. It implies a systematic and periodical monitoring and control of the implementation of the strategy, with information collected from the bottom and passed on to the top of the company.

Table 6.1 Detailed comparison between deliberate/rational and emergent strategy formulations

	Deliberate/rational	Emergent
Process	Pre-determined, based on rational models	Not decided, random
Activation	Systematic and periodical	Unsystematic
Competences	Clear distinction (top management decides, operatives apply)	Unclear distinction
Temporal sequence	First strategy formulation, later implementation	First solutions to local problems that later become "the strategy"
Direction	Top-down	Bottom-up
Knowledge of internal/ external environment	Assumed to be sufficient	Insufficient

Conversely, emergent strategy formulation assumes that little knowledge can be available in advance on markets, competitors, new technologies, and even internal components of the company. In the best case, strategists can learn by the experience of their operatives that face and solve local problem solving problems in a trial-and-error process. The decisions and courses of action that appear to provide better results emerge from bottom to top, are noticed and selected, and, maybe, can be later declared as "the company strategy".

Clearly, as Mintzberg and Waters (1985) also underline, a purely emergent approach may be paradoxical in real life, and it is difficult to imagine that companies can just work randomly with no direction at all. So, the usefulness of this abstract definition is that it helps to analyze when a particular situation is (more or less) close to an emergent strategy formulation: this can simply happen at some time of a company's life, or for special cases of strategic decision, or for specific parts of the company. Whatever it is, an important message here is that not necessarily a successful strategy must be decided by the top management with a rational approach, but can also come from local learning activities. A strategy is a complex and dynamic process which may require an appropriate combination of two different components: a deliberate component, designed according to the best knowledge available at the initial moment of its formulation, and an emergent strategy, which results as a necessary process of adaptation to the changeable business environment, and learning.

6.1.4 Multiple and Parallel Formulations

The idea of deliberate and rational strategy brings about an important implication in general terms: that decision making is a deterministic process (Bourgeois 1984). An important implication is that the decisions taken by company strategists, given the same initial conditions, should follow the same pattern and come to the same conclusion. This intrinsically normative or prescriptive approach is perfectly in line with the tradition of "one best way" (i.e.: "there is an optimal solution given certain conditions"), which went through the entire history of management, from Taylor's scientific management to World Class Manufacturing methods.

Here, an analogy with the history of physics can be useful. In Newtonian classical physics, the assumption is that there is linearity and determinism in phenomena (Bratianu 2007). So, if the laws describing phenomena are sufficiently valid, effects can be predicted when causes are known: under an Aristotelic perspective, we may say that consequences derive directly from premises. The same is for rational strategy formulation: we may say that the deterministic hypothesis makes a rational planning of future actions possible, once enough knowledge of the system under control is collected. So, if we know enough about a company and its competitive environment, we also know what to do: the rational models will bring us to a deterministic conclusion about what should be done. As clearly affirmed by Smith et al. (2010, p. 449), "Scholars have traditionally argued that organizational success depends on taking an 'either/or' approach to choosing between such paradoxical agendas: leaders assess the external environment, decide which agenda to favor, and then build a business model to implement this single, focused strategy. According to this view, success depends on proper alignment, both of the business model's internal aspects, and between it and the external environment."

But in the real life, it appears that the course of actions of companies is not so linear and unambiguous, and the "either/or approach to strategic tensions is inadequate" and "success over time is rooted in adopting this kind of 'both/and' approach, committing to paradoxical strategies and their associated product, market and organizational architecture" (Smith et al. 2010, p. 449). Notions such as the "ambidextrous organization" (Duncan 1976) suggest that companies may pursue completely different and maybe even conflicting strategies at the same time, because they can't find a "linear path" that bring them to an ultimate clear target.

Again, a parallel with physics is illuminating. A quantum-mechanics-like perspective leads to accept the unavoidable insufficiency of knowledge, and the idea that a system can be in different contrasting states at the same time. Similarly, in a "quantum-mechanics-like" economy, rational planning becomes impossible or at least useless (Mintzberg 1988; Nonaka and Zhu 2012; Spender 2014). Due to the intrinsic and irreducible uncertainty about the state of the system (the company and/or markets), companies may seem to follow different strategies at the same time (Miller and Friesen 1984) to face dynamic variations or insufficient knowledge; or while they, sometimes, appear to proceed by incremental steps in their course of action, in other cases, they change their plans to seek new stability. In addition, the awareness it is not possible to have complete and measurable information of the company nor of the competitive environment may force strategies to give importance to learning processes and lessons from experience. This also implies a change in mental capacities and cognitive attitudes (McKenzie et al. 2009): a capability to manage conflicting factors, and to bear the uncertainty and indeterminateness of ambiguous situations. The willingness to learn from experience (Beer et al. 2005), the openness to combine elements coming from the practice at a micro-level (Whittington 1996), and a consideration of the role of creativity (Bratianu 2007) become essential ingredients of strategic planning.

6.1.5 Planning Levels

Another point that complicates the analysis of strategy formulation is that all companies (and not only the big corporations) have a complex organization, often represented, in the management literature, by means of different *levels*.

Anthony (1965), for example, used to model the organizations by means of triangles or pyramids, that represent a hierarchical view of the management structure: operational decisions are placed at the bottom, intermediate (tactical) decisions in the middle, and strategic decision at the top (Fig. 6.3). Generally speaking, operational decisions are characterized by a higher degree of repetitiveness, standardization, and short-term horizon. Conversely, decisions at the top level of the company are characterized by frequent variations and long-term horizon.

This differentiation is significant in terms of strategy formulation as well. For example, in a company where operational "short term" activities prevail in terms of importance and value added (case A of Fig. 6.3), the overall strategy formulation of the company has a prevailing short-term orientation: there is less need to "see beyond", in direction to a distant future, and the knowledge required to formulate plans and future actions is less affected by uncertainty. In a company like case B, instead, the weight of "strategic decisions" is predominant: this means that the long term future of the company must be envisaged and planned, which is also challenging because it requires to face uncertainties and difficulties in managing the required knowledge.

Another version of hierarchical levels in strategy formulation is recalled by the typical distinction between corporate and business strategies (and related notions—

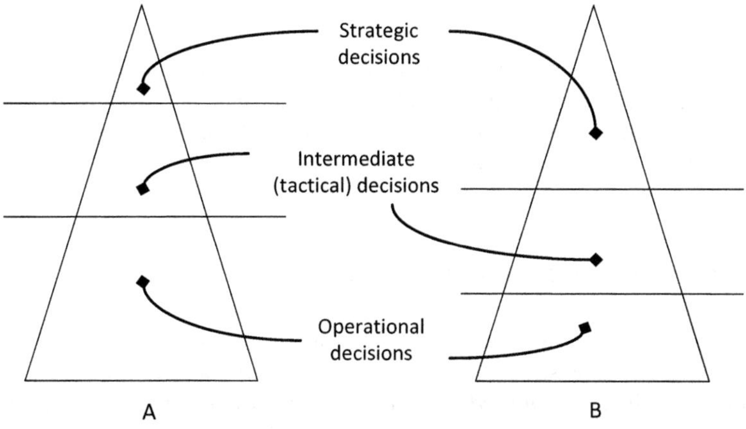

Fig. 6.3 Anthony's representation of hierarchical management levels

see e.g. Roach and Allen 1983), where the former represents the determination of the overall mix of business activities that are expected to rise the value of stakeholders, and the latter refers to specific goals and plans about products, productions, markets, etc. Considering Anthony's triangle of Fig. 6.3, we may say that corporate strategies are "more close to the top" of the triangle, while business strategies are near to the "bottom" operational activities. But the two strategies are, at least in principle, hierarchically connected to one another (i.e. the business strategy should be a detailed development of the general goals and plans of the corporate strategy). Also, this subdivision represents a typical engineering-based approach of *decomposing a problem* into more simple problems. At the higher level (namely, corporate strategy), decision makers need generic and general knowledge of the economic environment and of the expectations of stakeholders, while at a lower level (i.e. business strategy) decision makers must acquire more specific (and precise) knowledge to formulate plans and anticipate future actions.

6.1.6 Attempt of a Synthesis

So far, we have recalled some of the key elements of strategy formulation in business. We have seen that, both conceptually and practically, this process is far from being simple. Also, it is quite difficult to sketch its ultimate content, goal, or way of implementation.

As we have seen, a deliberate and rational approach is still an important reference for executives, that can use established models and schemes that help them to collect and analyze data, and to take decisions. But we have also considered that strategies may also emerge from the ground, i.e. from daily activities of operatives, whose local solutions can afterwards become "the strategy". In addition, strategic decisions are taken at different levels, and this requires different capabilities and knowledge contents.

Finally, we have learnt that, even in the same company, a turbulent environment and a condition of uncertainty may lead to different, sometimes conflicting, but coexisting strategies or strategy formulation approaches. In our view, this is probably one of the crucial challenges in today's management. Here, we will see if and how the notion of knowledge strategy can be of help here.

6.2 Knowledge Strategy Formulation

The analysis conducted before, regarding strategy formulation in general, can be applied to knowledge strategy as well. So, we will define knowledge strategy formulation as the *approach adopted to formulate a knowledge strategy*, just as it is done for the general strategy of a company. While the literature on knowledge strategy is rich, knowledge management (KM) scholars have devoted little attention

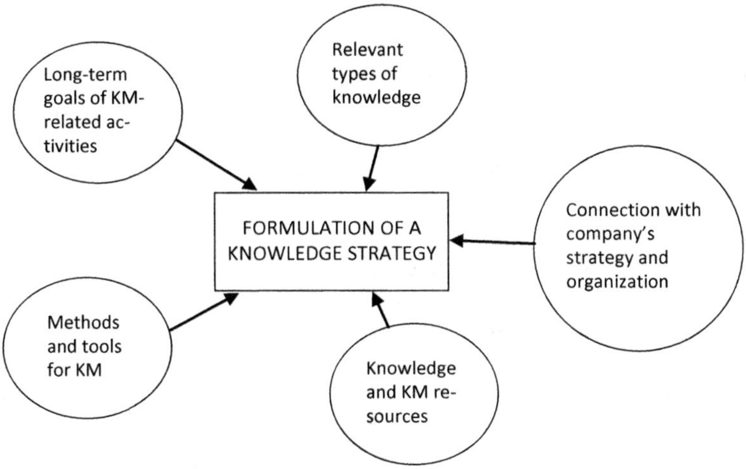

Fig. 6.4 Knowledge strategy formulation

to the process of knowledge strategy formulation, with few exceptions (see e.g., Holsapple and Jones 2006).

Formulating a knowledge strategy (Fig. 6.4) implies defining essential aspects, many of which recall the definition itself of knowledge strategy (Kim et al. 2003; Bolisani et al. 2014): the long-term goals of cognitive and learning processes, and the KM programs (i.e., what and how knowledge should be accessed, processed, and exploited, what value can be extracted from knowledge, etc.); the kinds of knowledge to be treated in the company; the methods and tools for achieving KM-related goals; the resources that must be adopted in KM programs. In addition, as we have learnt in the previous chapters, a knowledge strategy also implies a connection of the way knowledge is seen in the company, which means a link with the overall strategy of the company and also with its organizational structure.

Companies may adopt different approaches for deciding on these elements. Here, we can replicate the same analysis we made about strategy in general in the previous sections. In particular, we can apply the same distinction between deliberate/rational and emergent approach to strategy formulation (Donate and Canales 2012; Ichijo 2007; Snyman and Kruger 2004; Bolisani et al. 2016).

As mentioned (see previous sections), the *deliberate or rational* approach to strategy formulation planning is an explicit and rational formulation of goals, plans, and means that originates from precise intentions of the company, while an *emergent* approach refers to a situation where a strategy, rather than being defined in advance, can be better seen as an "ex-post" rationalization of a company's behavior. The main factor that can influence the position of a company is uncertainty: The more unpredictable and uncontrollable the internal or external environment, the more difficult the adoption of a rational strategic planning.

The point here is how these concepts can be used in the practice when it comes to knowledge strategies. Here, it is also useful to replicate a *hierarchical view* of strategy formulation, just like we suggested in Sect. 6.1.5. We can distinguish between *knowledge strategies* and *KM strategies* (see Chap. 5), and consequently we will have strategic formulation split into two steps: a) formulation of a *knowledge strategy* in strict sense (i.e., definition of the general view of knowledge as competitive resource of a company), and b) formulation of a *KM strategy* (namely, practical tools and methods through which knowledge is managed).

Considering these elements, a rational approach to knowledge strategy formulation can be intended as an approach where the general orientation of the company and the views of knowledge as strategic resource are deliberately designed at a top management level (Bolisani et al. 2016). While KM goals, methods, and tools are then formulated on the basis of the knowledge strategy previously defined and on a rational analysis of company's needs, objectives, and resources. Implementation and investment in detailed KM plans or methods consequently follow. Based on a rational analysis of internal resources and external competitive environment, the company decides which knowledge is important, what sources can be employed to develop this knowledge, etc. Later, the company decides and plans the practical methods and tools to implement that knowledge strategy, or the KM strategy is formulated on the basis of the goals and the framework defined by the knowledge strategy (Fig. 6.5).

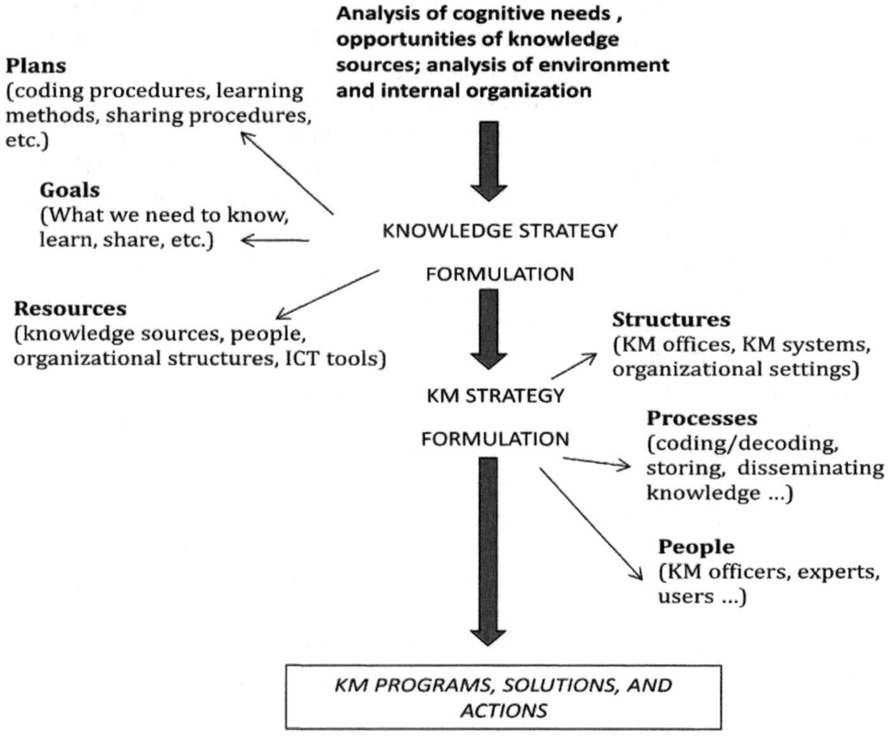

Fig. 6.5 Deliberate/rational approach to knowledge strategy formulation

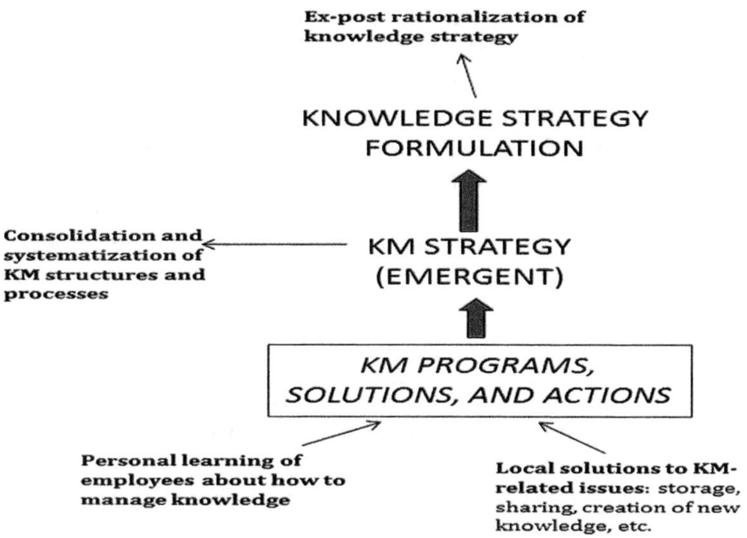

Fig. 6.6 Emergent approach to knowledge strategy formulation

The opposing view is that of an *emergent* approach to knowledge strategy formulation (Fig. 6.6). Here, practices, tools, and methods of KM originate from the daily practices and learning processes of a company's employees. In substance, employees research, develop and adopt their own methods of learning, storing, retrieving, and sharing knowledge in relation to their actual needs and practical problems to solve in their daily work. These methods and tools, when they prove to be effective, useful, and/or compatible with the daily business practices, become systematic routines and established practices and can be therefore recognized as "The KM Strategy". Afterwards, a process of rationalization can help to understand what this implies in terms of a "knowledge-based vision" of the company: in other words, a conceptual formulation of "The Knowledge Strategy" of the company becomes possible.

The existence of both approaches of deliberate and emergent knowledge strategy formulation have been recently recognized in the literature (Kotter 2012; Nonaka and Zhu 2012; Bolisani et al. 2016). Of course, these are generic categories and, in the real world, many variations and combinations are possible.

6.3 Knowledge and Strategy Formulation

What we have argued about knowledge and strategy formulation in this and in the previous chapters is that there are two key dimensions to consider. First, there are two extreme approaches to strategy formulation: one is called deliberate/rational, and it assumes a pre-defined model of the company and the environment that guides the processes of analysis and decision making (in a top-down logic); the other is called emergent, and it mostly rests on the local creativity of operative people whose solutions to specific problems acquire the status of an explicit strategy for the company with an ex-post rationalization (in a bottom-up logic).

Second, we have split the general approach to strategy in two elements: the company strategy and the knowledge strategy. These are strictly connected, but companies can ascribe a different degree of priority to one or the other.

In the following sections, our attempt is to connect these two dimensions, so that different patterns and paths that lead companies to strategic formulation can be singled out. We will then investigate the conditions that can lead companies to prefer one of these patterns.

6.3.1 Paths and Patterns in Strategy Formulation

Here, we will use and develop the analysis of the evidence of previous empirical research on strategic orientation. In Bolisani and Scarso (2015) four cases of companies were analyzed, and their approach to strategy and knowledge strategy was studied. Here, these four approaches will be re-classified by using the two dimensions considered earlier in this Sect. 6.3. Of course, these are only four cases, so they will be simply used as paradigmatic examples, but with no meaning of generalization: indeed, other companies (and even the same ones that we analyze here) may follow different combinations and variations of the paths to strategic formulation that we describe here.

In addition, not necessarily the companies use the same terms that we adopt here in this book to define and classify strategies. What is important for our purposes is the description of what they do. Finally, the cases presented here are referred to a specific time, so strategies can be changed from then.

Case 1. Oil Company

This is a big oil multinational with business units and operational locations scattered in 70 different countries, more than 75,000 employees, and about 150 billion sales. The company has a structure for strategic planning that operates quite formally and on a regular basis, with plans formulated on a time horizon of 4 years, and split from corporate level to the different business units (Cunningham and Harney 2012). Company strategists are used to monitoring markets and competitive environment on a regular basis and, based on this analysis, to formulate the company strategy.

From about the beginning of the new Millennium, the company was in a difficult economic climate, with a rapidly decreasing oil price and an increasing concentration in the industry, with the big players growing fast by means of M&A and improving their market power. Being not possible to grow internally any more, for the company, it became essential a strategy of fast internationalization. In particular, a strategy of research for new reservoirs, acquisition and management of those possessed by others was therefore decided.

However, executives were also aware of the increasing technical difficulties of exploration and production of oil and gas. Reservoirs became more and more difficult to find and exploit, and this required huge investments and sophisticated technologies. This adds to the usual complexity of the oil business, that is based on many different activities and professional skills (not only geology, chemistry, construction, electronics and mechanics, but also economic-managerial capabilities and legal–political competencies for negotiation and contracting) that must be integrated to one another. Furthermore, oil companies must combine routine processes (for instance, day-to-day oil extraction—that require efficiency and standardization), with project-based activities (e.g. discovery of new reservoirs—that involve solutions to specific and partly new problems). In this environment, any wrong decision taken by professionals working at an exploration site or well, may cause huge losses to the firm and can weaken its position.

The rational analysis of this situations led executives to conclude that, to implement the strategy that the company had decided to adopt, there was a problem of knowledge, related to three main aspects: a) the necessity to make young professionals learn faster by eliminating spatial and time constraints, so that they could more easily take appropriate decisions when they work at some production site somewhere in the world, by exploiting the advice and past experience of senior professionals in the company, whenever they were working; b) the need to build a sort of virtuous cycle in knowledge creation and re-use: in other words, decisions taken on the ground should have been reported for future reuse by others; and c) the need to integrate the different elements of knowledge by creating an environment that could facilitate sharing. All this was considered an essential ingredient of the knowledge strategy.

Consequently, a knowledge strategy was formulated; particularly, it was decided to build an online environment that facilitates advice and knowledge exchange between the professionals working at some local production site and the more experienced people that can be wherever around the world. The KM system that was set up now consisted of a group of 11 online communities of practice, specializing in different areas (e.g. geology, chemistry, facility management, etc.). Each community is formed by experienced senior professionals who are connected by means of an internal e-mailing system. The community is activated, on demand, when a professional needs some help, for instance because there is a problem at some oil well. The experts propose a possible solution based on their experience, and can therefore transfer their knowledge to the younger colleagues by means of the system. This makes the responsiveness of the entire company higher, and reduces the risk of losses or mistakes thanks to a better circulation of knowledge across the company.

Case 2. Consulting Company

This is a major global consulting company, leader in assurance, tax, transaction and advisory services. As a KIBS firm, its business is based entirely on the knowledge of its 170,000 professionals, who are dispersed all over the world. In a company of this scale, the formulation of a knowledge strategy is critical and challenging. There is the need to operate in a variety of sectors, with many clients, and with diverse business units, each one characterized by a specific knowledge base. Particularly, executives are aware that the crucial element of exchanging, sharing and reusing knowledge within the organization at a global scale is crucial to deliver more value to clients, and to act faster. Knowledge is a resource that must not be confined to a particular area but has to be shared among all the hundreds of offices and professionals worldwide. The essential element of the knowledge strategy is that it is necessary not only to share explicit knowledge but also to capture and retain valuable tacit knowledge that resides in the minds of employees. In other words, there is the need to map, connect and enhance all the different knowledge resources (i.e. distinct offices, professionals and databases) that are scattered worldwide. In 2007, the knowledge strategy of the company was formulated in terms that the goal was to design and implement a single global knowledge function across the globe.

As a consequence, the KM strategy of the Company, that defines the specific KM offices, functions and plans, has been formulated with the explicit goal to enhance the company's capacity to create and share knowledge resources globally, and to improve the learning capability of service offices or individual professionals. In particular, the KM plan consists of the definition of a common "knowledge platform" with the aim to define shared taxonomies for the entire company (i.e. only one way to call things), a "one-stop shop" system to search and find appropriate pieces of knowledge for each single professional or office, and a centralized maintenance process, which saves time and money at the local level. The actual project started in early 2009 with a first release of the new platform later that year.

In light of this knowledge strategy, the company strategy also improved. The Advisory business is based on addressing big and complex industry issues to help clients grow, optimize and protect their business. So, by means of the KM platform, the diverse advisory teams, supported by the global connectivity platform, have the possibility to react faster to the requests of their clients, and can activate a more effective response and improved service quality. This represents an element of the competitive strategy for the company.

Case 3. Electronic Company

This company is a holding of nine small firms located in Italy and the USA, with about 400 employees and a 70 million euros annual turnover. The company produces small personal computer devices for special purposes. Since 2001, the company exploited the opportunity to acquire other innovative small companies, which has led to an expansion of the business, with new product lines and new clients to manage. This resulted in a new opportunity, for the company, to keep the pace of leading edge technologies of electronic miniaturization and reinforce competitiveness. However, it also requires an effort to put order to this fast growing organization, and a continuous improvement and maintenance of the knowledge base. In addition, and as a consequence of this business growth, new knowledge was acquired by the company. New and old employees, therefore, needed integration and re-combination of their different languages, approaches and perspectives that they were used to employ locally in their daily business.

All this was not achieved by means of deliberate KM plans but, rather, of other actions targeted to different aims. However, the top management progressively gained awareness that a more systematic approach to the management of knowledge could have been useful. In fact, on the one hand, the Company had to favor the circulation of knowledge in the entire organization, so that each distinct part can easily react to the challenges of the competitive environment; on the other hand, it had to protect the distinctive competences that have enabled it to attain a leading market position.

In light of this, a definition of KM plans was made, to manage the different kinds of knowledge in the company. On the basis of the experience of employees, a codification process was activated: the new elements of knowledge that are acquired from outside are codified (whenever possible) in a standard format, so that they can be accessed to all the distinct parts of the organization. Central repositories such as MS SharePoint are also used. Instead, the core distinctive elements of knowledge of the company remain tacit and embedded in the minds of expert employees: these expert people are, however, now temporarily assigned "on demand" to some specific part of the organization, when there is the need to exploit their capabilities.

Case 4. IT Services

This is a small IT services company located in Northern Italy with 30 employees and 5 external collaborators; the annual turnover is about 5 million euros. The firm provides custom-made IT platforms to clients, and is a project-led company: a project for a client consists of various steps (first contact, needs analysis, negotiation, technical implementation, delivery and customer support).

These different steps involve distinct departments of the organization (i.e. sales force, technical staff and customer service) that represent knowledge resources of the organization that are distinct but must be connected and shared to be exploited effectively. For example, to provide effective assistance to a customer, the support department must rapidly retrieve all the knowledge about the installed system (e.g. maps, codes, network configurations and functioning parameters), but this knowledge was produced by others (the technical staff) at another time and by using different approaches. Similarly, when the sales agents have to formulate a new commercial proposal, they may be benefited from retrieving the technical elements of past installations.

Initially, when the company was founded, this problem already existed but was easily resolved because there were few employees, few customers and few projects: direct face-to-face contact and knowledge exchange were possible. With the growth of the company, the management started to introduce some mechanisms and tools to store and retrieve project documentations. However, this was made without a systematic reflection on the KM needs of the company and on the possible solutions to these. In fact, the company initially used simple paper-based documentation, which proved to be ineffective; later, a shared folder was used, which was more efficient but inflexible. Recently, an internal wiki system has been implemented that contains all the relevant information about any installation of a new system for a customer. The adoption of the system was not the consequence of a planned strategy but a solution that has emerged progressively. As well underlined by one of its creators, the approach followed in the development of the system "was contingent and closely tied to the tools available". It is only now that the wiki system has proven to be useful and used, and its further development is now part of the deliberate KM plans of the company.

The use of this tool is now becoming essential for the business of the company. It allows to improve the quality and effectiveness of the maintenance service, and makes it possible to pursue a company strategy to loyalize clients. This proves to be an essential competitive weapon, especially in turbulent times.

6.3.2 Analysis

As we mentioned, it is not possible to generalize from a handful of cases. In addition, ours is an "ex-post" analysis, that made it possible to identify and single out specific patterns in strategic formulation, regardless that the single companies are aware of that. In any case, what we propose here is the examination of the four cases previously described, by adopting the perspectives developed in the earlier chapters, and with the special purpose to single out different paths in strategic formulation.

The first two cases (1 and 2) show some similarities in the way strategies are formulated. Both companies apparently adopt a deliberate/rational approach: first, an analysis of problems, needs or resources, and next a formulation of the strategy. Their approach appears to follow a top-down logic: first, definition of strategies at a corporate level, and later details at operative level (business strategy and KM strategy).

However, they seem to follow a different path. In Case 1 (Oil company), there was first a consideration and assessment of market needs, competitive pressures and position, and external business opportunities (namely: what will be our markets? What are our competitors doing? Etc.). On this basis, it later emerged that there was "a problem of knowledge" and "of KM". So, we may say that, for this company, the priority is that of the "company strategy". Conversely, in Case 2 (Consulting company), the starting point of strategy formulation was the assessment of internal knowledge resources and how these can be mobilized and improved (i.e.: what do we know? How do we know and how do we manage this knowledge?). Therefore, in this approach, there is a special attention for knowledge strategy, which also influences the formulation of the company strategy (that defines the business). Similarly, if we compare Case 3 and Case 4, while both start from problems emerging "from the ground", in Case 3 the company first focused on the business opportunities emerged from a M&A policy, and the consequent problems of KM that later emerged. In Case 4, the company was trying to solve specific problems in knowledge exchange, that then became established solutions and, next, were also seen as potential business opportunities.

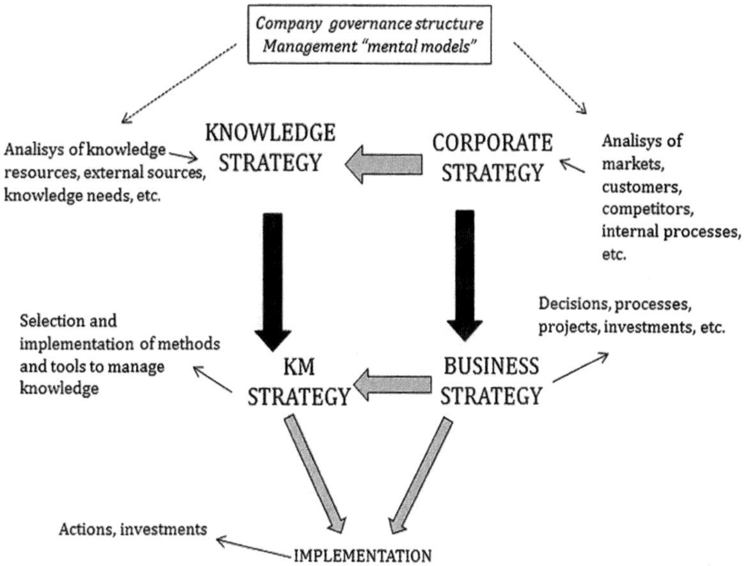

Fig. 6.7 Deliberate/rational strategy formulation with an emphasis on company strategy

So, there are four different paths in strategy formulation that can be summarized as in Figs. 6.7, 6.8, 6.9 and 6.10. Figure 6.7 depicts the case of *"deliberate/rational strategy formulation where the emphasis is on company strategy"* (Case 1 in our analysis). So, strategists first focus "on the business" and, following a top-down logic, formulate the corporate strategy (and later the business strategy). Knowledge and KM strategies are formulated as a consequence of corporate and business strategies.

Figure 6.8 (Case 2 of our analysis) is partially a reversed situation, where there is still a deliberate/rational approach and a top-down logic, but an emphasis is put on knowledge resources first. So, knowledge strategy and, late, KM strategy influence the formulation of corporate and business strategies (i.e. the company strategy). Figures 6.7 and 6.8 also emphasize that strategy formulation is based on a sort of "pre-defined" mental environment that characterizes how strategists conduct their analysis and take their decisions. Mental models, capabilities and education background, values, but also governance style and structure of the company guide the way strategists operate.

Figures 6.9 and 6.10 illustrate the cases of an emergent strategy formulation. In Fig. 6.9 (Case 3 of our analysis), the emphasis is first on "the business": solutions to local business problems progressively become "the strategy" of the company, in a bottom-up logic, and also influence the KM solutions and strategies.

Figure 6.10 (Case 4 of our analysis) refers to the case where it is the local solutions of employees to their specific KM problems that become an established KM and knowledge strategy. Later, these also become business opportunities and therefore influence the perception and the formulation of the company strategy.

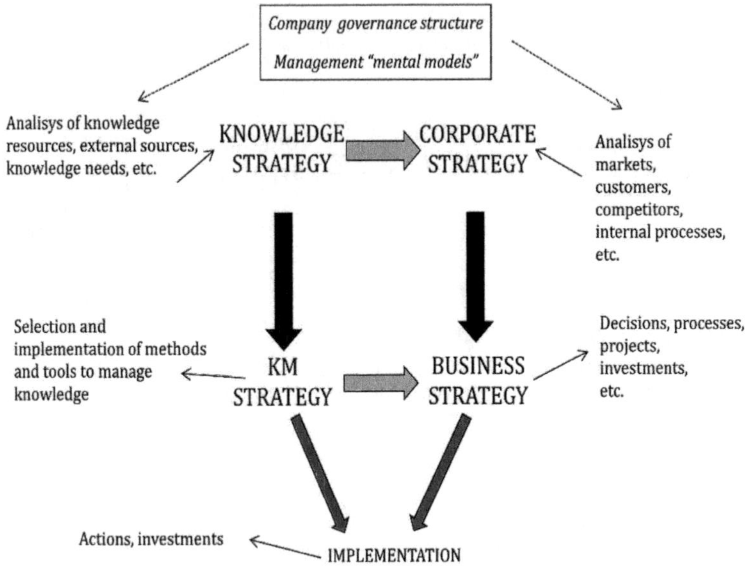

Fig. 6.8 Deliberate/rational strategy formulation with an emphasis on knowledge strategy

Fig. 6.9 Emergent strategy formulation with an emphasis on company strategy

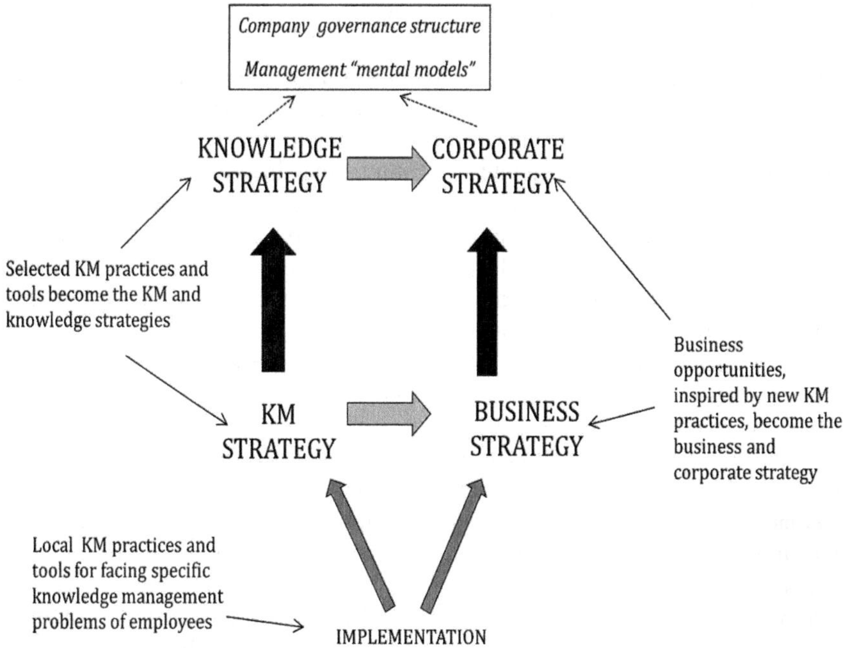

Fig. 6.10 Emergent strategy formulation with an emphasis on knowledge strategy

6.4 An Integrated Model: Formulating Strategies in the Learning Organization

What are the conditions that can lead to a particular path of strategy formulation? What are the pros and cons of each approach? As regards the first question, an important aspect can be the nature and characteristics of the business: for example, the big size of a company may imply a more formal and hierarchical approach to strategy formulation, and a deliberate/rational approach and a top-down logic can be a consistent application of this governance style (cases 1 and 2 of the previous section can be good examples of this). Conversely, small companies may have a rather informal way of adopting a strategy (i.e. cases 3 and 4), and an emergent approach can be more likely. In addition, as we have already mentioned, when the key elements of the business are represented by intangibles and, more precisely, knowledge (as in the case of KIBS companies), this can signal a priority on knowledge when strategies are formulated (i.e. the business is the consequence of the knowledge that a company manages); cases 2 and 4 are good examples of this. Instead, when tangible operational activities prevail (like in traditional manufacturers—cases 1 and 3 above), the formulation of a strategy starts from "the business side" and later extends to the knowledge resources. Clearly, there may be other elements that can be considered, which can also be the object of further research that, on this topic, is still scarce.

A second key question is the pros and cons of the different strategic paths under the various conditions where a company operates. In particular, a central point of our analysis is the irreducible uncertainty of the internal and external environment, and the turbulence or unpredictable variability that increasingly characterize all economic activities. The negative effects of uncertainty and turbulence on decision-making processes can depend on intrinsic and objective elements (i.e.: the environment is intrinsically complex, and we can't achieve sufficiently reliable knowledge regardless the efforts that we make) or on subjective elements (namely: the knowledge processing capability of strategists).

We can associate the degree of uncertainty and turbulence—and the incapability to manage them—to the strategy formulation path that the companies can adopt. As we have mentioned in the previous chapters, in a stable and predictable environment, strategists are more confident that pre-defined decision models and rational schemes can be successful. Also, big companies that have a formal structure and governance style, or that can invest in people directly specializing in "strategizing", can be more confident about their capability to collect and analyze information, and process it into knowledge for better strategy formulation.

Conversely, when the environment is instable and turbulent, or too complex to analyze (in absolute terms, or in relation to the cognitive capabilities of managers) the deliberate/rational approach may be not applicable, or may lead to unjustified trust in rationality and consequent failures. Under these conditions, an emergent approach to strategy formulation means that there is no intention or possibility to rely on pre-defined rational schemes and mental models, and the hope is that strategies simply emerge from the effective solutions that are applied on the ground. "The basic argument here is that when humans must make decisions in situations in which causality is poorly understood, where there is considerable uncertainty, where people hold different beliefs, where they have personal biases, where they do not understand each other and where they lack all the required technical expertise, then decisions are made and actions taken on an irrational basis" (Stacey et al. 2000, p. 169). In principle, this should allow better adaptation to turbulence and unexpected or unpredictable events: only the local solutions that prove to be effective and successful are selected and become established as a strategy. An emergent approach assumes it is impossible to make a rational analysis, because uncertainty is too high, and the environment is too turbulent and dynamic: The company can simply adapt its practices to the emerging needs and opportunities over time. Formal definitions of these KM practices as a knowledge strategy become an ex-post exercise. Turbulences in the external environment may bring the company on the "edge of chaos." "In an intermediate state, between stability and instability, the dynamics known as 'the edge of chaos' occurs; namely, the paradox of stable instability. … At the edge of chaos, novelty emerges in a radically unpredictable way" (Stacey et al. 2000, p. 112). The concept of emergent strategy "opens the door to strategic learning, because it acknowledges the organization's capacity to experiment" (Mintzberg 2000, p. 189).

The problem of an emergent approach is, at least in its extreme manifestations, it is apparently a pure "wait-and-see" behavior: The company simply adopts the practices that, occasionally and unpredictably, emerge from the ground and from the concrete experience of employees, but with no effort of anticipation. Also, in an abstract sense, emergent strategy formulations imply that decision makers don't have any pre-defined mental model, system of value, or decisional structure, which is hard to imagine in any human organization: humans can't simply live "in a vacuum", and need at least some basic rational references to decide and act. Even the use of the simplest language implies a "form of thought" that eventually influences the patterns of decision-making.

Other differences can be underlined between companies that have an emphasis on knowledge strategies and those that see a priority for company strategies. Using the terminology of knowledge management, we may expect that privileging the formulation of company strategy implies a focus on the explicit and objective forms of knowledge: it is assumed that the environment is knowledgeable in objective terms, and it is just a matter of applying the appropriate rational models. Instead, an emphasis on knowledge strategy means a focus on the whole range of dimensions that knowledge can have. Therefore, in formulating a strategy, strategists can consider questions e.g.: what do we don't know that we would really like to learn? What do we consider important, interesting, fascinating to learn, or simply closer to our emotional or spiritual sphere? How do we like to perform our learning processes? How do we aim to socialize our learning processes? It is not that a focus on company strategy is free from these "subjective" elements, but generally speaking they are considered a problem to avoid (in the seek for objectivity and repeatability in decision-making), and not a value to exploit.

In summary, all the different possible orientations in strategy formulation can have advantages and disadvantages, in relation to the characteristics of the environment, the degree of uncertainty and turbulence, the management style, and the social, cultural, emotional or spiritual values in the company. In this section, we develop the idea of an *integrated approach* that aims to exploit the advantages of the distinct approaches. This *integrated view* sees strategy formulation as a continuous effort of *learning* (Fig. 6.11) with an orientation to strategic decisions that combines a top-down and bottom-up logic.

So, strategists may combine a provisional formulation of a general vision on the basis of *possible scenarios* that apparently fit the current understanding of the environment and the ideas of what the company and its future may or should be, on the basis of pre-existing mental models of the world, management styles, and governance structures. However, the definition of these strategies also comes from the needs and opportunities that dynamically emerge from the ground and the daily practices. A dynamic interaction between the lower-and-upper and the right-and-left parts of the scheme has advantages. For a company, it is possible to formulate "preliminary strategies" based on a rational analysis of the current and future situation, which provides an initial reference for implementing practices and for setting up possible research paths of new knowledge and mechanisms of individual or organizational learning. Here, the reference to generic knowledge strategies can

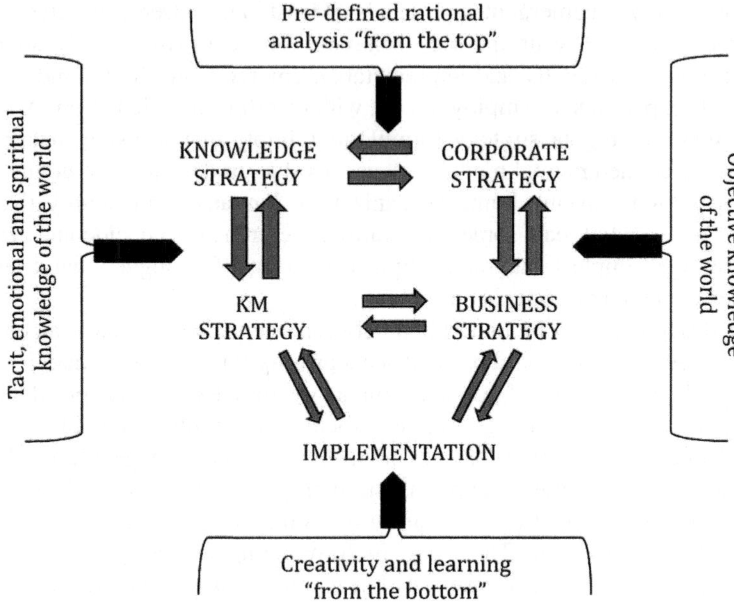

Fig. 6.11 An integrated approach to strategy formulation

be of use. This avoids operative employees simply to "go blindly" or adopt their KM practices on the basis of occasional opportunities. At the same time, top management must keep their antennas up to get everything useful that comes from the ground: The results of individual learning processes can provide novel ideas or solutions to unexpected problems. More precisely, in KM, this implies the capability to absorb, analyze, and rationalize the interesting spontaneous practices that are "invented" and applied by single employees and to modify the general knowledge strategy if so required.

At the same time, strategist may combine a right-to-left and a left-to-right path: on the one hand, by adopting an "objective view" of knowledge based on the vision of the "actual business", they will understand the business needs and formulating the business goals and, based on that, they will see what the company already knows and what is left to be known. But at the same time, they will be aware that this objective view is just based on a more general vision of what the company likes to know and learn, so there will be space for emotions and spiritual values that will, in turn, influence the way the world will be envisaged.

To sum up, strategy formulation becomes a complex process of generative learning (Mintzberg 2000; Senge et al. 1999) able to cope with the emergent future (Stacey et al. 2000). Decision making processes are an iterative and co-creating process that goes beyond the Newtonian logic. Senge et al. (2005, p. 90) use, for such a complex mental process, the concept of *presencing*: "In effect, presencing constitutes a third type of seeing, beyond seeing external reality and beyond even seeing from within the living whole. It is seeing from within the source from which the future whole is emerging, peering back at the present from the future." This integrated knowledge strategy planning may be not only a pure revision and adaptation of existing planning methods, but instead, may require new mental schemes and practical techniques. In our view (and wishes), this also can provide inspiration for new research on strategic thinking.

6.5 A Summary

In this chapter, we have discussed the key issues of strategy formulation, and we have argued that a deliberate and rational approach is, in principle, comfortable and comforting for executives, that can count on objective models and repeatable processes to take their decisions. Especially (but not only) in large corporations, the application of this "planning" approach has led to formal decision structures, that attempt to "work objectively", i.e. free from the influence of human subjectivity and value. In the literature, there is an established tradition of methods for formal strategic planning that is still important today.

However, we have also seen how the awareness of uncertainties and turbulences that characterize the context of decision making have challenged this objective vision of strategy formulation. From the pioneering works of Mintzberg and other colleagues, it has become clear that strategies can be also the ex-post rationalization of solutions that more or less emerge (even randomly) from the daily practice of operatives in companies. In principle, this gives space to creativity and spontaneous learning, and can make companies adapt more easily to unexpected changes in markets, technologies, and competitive environments.

Similarly, we have argued that strategy formulation is not simply the application of "neutral" models through which reality can be seen objectively: strategizing is indeed a cognitive process, but knowledge of human beings has several dimensions, that also include emotional aspects and spiritual values. So, even the "most neutral vision" or reality is affected by this background, and strategy formulation should also consider what a company can or wishes to know and learn.

To combine these different views, we proposed an integrated approach to strategy formulation. This approach combines the different possible orientations to strategy formulation, and adopts a view of companies as learning organizations where the single individual at operative level and the management structure must understand how to interact fruitfully, where rational models and creativity must coexist and mutually exploit one another, and where emotional and spiritual components of knowledge are not left apart but become part of a new way of thinking and seeing the possible or desired future.

References

Ansoff, H. I. (1965). *Corporate strategy: An analytic approach to business policy for growth and expansion*. New York: McGraw-Hill.

Anthony, R. N. (1965). *Planning and control: A framework for analysis*. Cambridge, MA: Harvard University Press.

Beer, M., Voelpel, S. C., Leibold, M., & Tekie, E. B. (2005). Strategic management as organizational learning: Developing fit and alignment through a disciplined process. *Long Range Planning, 38*(5), 445–465.

Bolisani, E., & Scarso, E. (2015). Strategic planning approaches to knowledge management: A taxonomy. *Vine – Journal of Information and Knowledge Management Systems, 45*(4), 495–508.

Bolisani, E., Paiola, M., & Scarso, E. (2014). Cognitive features of KIBS companies: Evidence from a cluster analysis. *International Journal of Services, Technology and Management, 20* (4–6), 215–232.

Bolisani, E., Scarso, E., & Zięba, M. (2016). How to deal with knowledge in small companies? Defining emergent KM approach. *International Journal of Learning and Intellectual Capital, 13*(2–3), 104–118.

Bourgeois, L. J. (1984). Strategic management and determinism. *The Academy of Management Review, 9*(4), 586–596.

Bourgeois, L. J., & Brodwin, D. R. (1984). Strategic implementation: Five approaches to an elusive phenomenon. *Strategic Management Journal, 5*(3), 241–264.

Bratianu, C. (2007, September 6–7). Thinking patterns and knowledge dynamics. In *Proceedings of the 8th European Conference on Knowledge Management, Barcelona* (pp. 152–157). Reading, MA: Academic Conferences Ltd.

Burgelman, R. A. (1983). A model of the interaction of strategic behavior, corporate context, and the concept of strategy. *The Academy of Management Review, 8*(1), 61–70.

Chandler, A. D. (1962). *Strategy and structure: Chapters in the history of the industrial enterprise*. Cambridge, MA: MIT Press.

Cunningham, J., & Harney, B. (2012). *Strategy and strategists*. Oxford: Oxford University Press.

Donate, M. J., & Canales, J. I. (2012). A new approach to the concept of knowledge strategy. *Journal of Knowledge Management, 16*(1), 22–44.

Drucker, P. F. (1954). *The practice of management*. New York: Harper & Brothers.

Duncan, R. (1976). The ambidextrous organization: Designing dual structures for innovation. In R. H. Killman, L. R. Pondy, & D. Sleven (Eds.), *The management of organization* (pp. 167–188). New York: North Holland.

Elfring, T., & Volberda, H. W. (2001). Schools of thought in strategic management: Fragmentation, integration or synthesis. In H. W. Volberda & T. Elfring (Eds.), *Rethinking strategy* (pp. 1–25). London: Sage.

Gluck, F. W., Kaufman, S. P., & Walleck, A. S. (1980). Strategic management for competitive advantage. *Harvard Business Review, 58*(4), 154–161.

Heracleous, L. (1998). Strategic thinking or strategic planning? *Long Range Planning, 31*(3), 481–487.

Holsapple, C. W., & Jones, K. (2006). Knowledge management strategy formation. In D. G. Schwartz (Ed.), *Encyclopedia of knowledge management* (pp. 419–428). Hershey, PA: Idea Group.

Hussey, D. E. (2007). *Strategic management: From theory to implementation*. Milton Park: Taylor & Francis.

Ichijo, K. (2007). The strategic management of knowledge. In K. Ichijo & I. Nonaka (Eds.), *Knowledge creation and management. New challenges for managers* (pp. 121–145). Oxford: Oxford University Press.

Karami, A. (2012). *Strategy formulation in entrepreneurial firms*. Aldershot: Ashgate.

Kim, Y.-G., Yu, S.-H., & Lee, J.-H. (2003). Knowledge strategy planning: Methodology and case. *Expert Systems with Applications, 24*(3), 295–307.

Kotter, J. P. (2012). Accelerate. *Harvard Business Review, November*, 45–58.

Martinet, A. C. (2010). Strategic planning, strategic management, strategic foresight: The seminal work of H. Igor Ansoff. *Technological Forecasting and Social Change, 77*(9), 1485–1487.

McKenzie, J., Woolf, N., & van Winkelen, C. (2009). Cognition in strategic decision making. A model of non-conventional thinking capacities for complex situations. *Management Decision, 47*(2), 209–232.

Miller, D., & Friesen, P. H. (1984). *Organizations: A quantum view*. Englewood Cliffs, NJ: Prentice-Hall.

Mintzberg, H. (1987). Crafting strategy. *The McKinsey Quarterly, Summer*, 71–90.

Mintzberg, H. (1988). Generic strategies: Toward a comprehensive framework. *Advances in Strategic Management, 5*(1), 1–67.

Mintzberg, H. (1990). Strategic formation: Schools of thought. In J. W. Frerdrickson (Ed.), *Perspectives on strategic management* (pp. 105–235). New York: Harper & Row.

Mintzberg, H. (1994). The fall and rise of strategic planning. *Harvard Business Review, 72*(1), 107–114.

Mintzberg, H. (2000). *The rise and fall of strategic planning*. London: Pearson Education.

Mintzberg, H., & Waters, J. A. (1985). Of strategies, deliberate and emergent. *Strategic Management Journal, 6*(3), 257–272.

Nonaka, I., & Zhu, Z. (2012). *Pragmatic strategy. Eastern wisdom, global success*. Cambridge: Cambridge University Press.

Ohmae, K. (1983). *The mind of the strategist*. Harmondsworth: Penguin.

Roach, J. D. C., & Allen, M. G. (1983). Strengthening the strategic management process. In K. J. Albert (Ed.), *The strategic management handbook* (pp. 197–207). New York: McGraw Hill.

Roth, A. V. (1996). Achieving strategic agility through economies of knowledge. *Planning Review, 24*(2), 30–36.

Senge, P., Kleiner, A., Ross, R., Roth, G., & Smith, B. (1999). *The dance of change*. New York: Currency Doubleday.

Senge, P., Scharmer, C. O., Jaworski, J., & Flowers, B. S. (2005). *Presence: Exploring profound change in people, organizations, and society*. New York: Currency Doubleday.

Simerson, B. K. (2011). *Strategic planning: A practical guide to strategy formulation and execution*. Santa Barbara, CA: ABC-CLIO.

Simon, H. A. (1960). *The new science of management decision*. New York: Harper and Row.

Smith, W. K., Binns, A., & Tushman, M. L. (2010). Complex business models: Managing strategic paradoxes simultaneously. *Long Range Planning, 43*(2), 448–461.

Snyman, R., & Kruger, C. J. (2004). The interdependency between strategic management and strategic knowledge management. *Journal of Knowledge Management, 8*(1), 5–19.

Spender, J. C. (2014). *Business strategy. Managing uncertainty, opportunity, & enterprise*. Oxford: Oxford University Press.

Stacey, R. D., Griffin, D., & Shaw, P. (2000). *Complexity and management: Fad or radical challenge to systems thinking?* London: Routledge.

Ulwick, A. (1999). *Business strategy formulation*. Westport, CT: Quorum Books.

Whittington, R. (1996). Strategy as practice. *Long Range Planning, 29*(5), 731–735.

Generic Knowledge Strategies

7

The purpose of this chapter is to introduce generic knowledge strategies, which aim at increasing the level of organizational knowledge and creating the intangible infrastructure for company strategies and achieving competitive advantage. The main characteristic of these generic strategies is that they can be developed in any organization although their success is related to a specific organizational context and a given business environment. The ontology of these generic strategies comes from the equilibrium dynamics of organizational knowledge and the correlation with the known-unknowns matrix. The generic knowledge strategies presented in this chapter are the following: exploitation strategies, acquisition strategies, sharing strategies and exploration or knowledge creation strategies. Exploitation knowledge strategies are designed in a similar way to low cost business strategies and efficiency models. This is a consequence of the fact that managers know what they know, which means that they know very well their intangible resources. Acquisition knowledge strategies are designed as a result of the identification of a strategic knowledge gap. Sharing knowledge strategies are specific for knowledge management and they contribute to increase the level of organizational knowledge by its diffusion within the whole organization. Exploration knowledge strategies focus on knowledge creation and on feeding the innovation process.

7.1 Thinking Perspectives

7.1.1 Generic Business Perspectives

According to *Oxford advanced learner's dictionary* "generic" means "shared by, including or typical of a whole group of things; not specific". Essentially, "generic strategies" can be developed by any organization in concordance with its vision and available resources and capabilities. From the previous chapters we learned how complex and diverse is the field of knowledge strategies and how compelling is to consider for further analysis a multidimensional framework. Three main

E. Bolisani, C. Bratianu, *Emergent Knowledge Strategies*, Knowledge Management and Organizational Learning 4, DOI 10.1007/978-3-319-60657-6_7

perspectives may be considered as possible initial dimensions of that framework: a) Porter's generic business strategies; b) the known-unknown matrix, and c) the organizational knowledge dynamics. Our conceptual research demonstrates that regardless of the initial starting point, the final result converges toward the same spectrum of generic knowledge strategies, even if the angles of their perception might be a little different.

The concept of *generic strategies* has been introduced by Michael Porter in his seminal book *Competitive advantage: Creating and sustaining superior performance* (1985). Focusing on the competitive position a firm might have, Porter considers that there are two fundamental approaches to competitive strategy: low cost and differentiation. The significance and evaluation of any strength or weakness a firm possesses is essentially a function of its impact on the cost level and differentiation. "The two basic types of competitive advantage combined with the scope of activities for which a firm seeks to achieve them lead to three generic strategies for achieving above-average performance in an industry: cost leadership, differentiation, and focus. The focus strategy has two variants, cost focus and differentiation focus" (Porter 1985, p. 11). The cost leadership and differentiation strategies are conceived for industry wide, while the focus strategy is conceived for only a segment of that industry. The firm that develops *a cost leadership* strategy aims to achieve competitive advantage through a series of efficiency methods which lead to the lowest cost per product or service on the market. The cost leadership strategy implies mass production and a large volume of products and services sold. These products have basic and functional features able to satisfy the customers from a very vital and practical perspective. That allows the firm to become an above-the-average performer in its industry, and as a consequence it will obtain higher returns than its below-the-average competitors. In contrast to cost leadership strategy, *the differentiation strategy* addresses people with new and different psychological needs. In developing this strategy, the firm "selects one or more attributes that many buyers in an industry perceive as important, and uniquely positions itself to meet those needs. It is rewarded for its uniqueness with a premium price" (Porter 1985, p. 14). Differentiation can be thought in product design, realization, or delivery system. Generating novelty and embedding it into firm's activities are the sources of any differentiation strategy. While cost leadership and differentiation strategies can be designed for a large industry range, *focus strategy* is designed only for a segment or several segments of that industry. "By optimizing its strategy for the target segments, the focuser seeks to achieve a competitive advantage in its target segments even though it does not possess a competitive advantage overall" (Porter 1985, p. 15).

To understand Michael Porter's vision about strategy we must add his remark that operational effectiveness is not a strategy. Although it contributes to realize a cost leadership strategy, operational effectiveness is based on short time thinking and decision making without any direct impact on competitive strategy. Only through an integration process and convergence pattern of thinking toward competition and competitors a firm can achieve a sustainable competitive advantage by "performing different activities from rivals' or performing similar activities in different ways" (Porter 1996, p. 3). That is valid when we consider both tangible and intangible resources, although they have different behavior and ways of being processed. When we map the cost leadership strategy from the tangible resources domain onto the intangible resources domain we get what March (1991) and Zack (1999) called *exploitation strategy*. Exploitation of organizational knowledge means a good understanding of what does exist at a certain moment within organization as cognitive, emotional, and spiritual knowledge, in explicit and tacit forms. Also, it is about knowing and using efficiently data, information, and knowledge stored in the information systems. Exploitation stimulates knowledge codification, sharing, dissemination, propagation and embedding. Exploitation means to reduce knowledge waste and knowledge loss, by increasing knowledge retention and knowledge reuse. When we map the differentiation strategy from the tangible resources domain onto the intangible resources domain, we get the *exploration strategy* (March 1991; Zack 1999). Exploration means to search for new knowledge and ways of increasing the level of organizational knowledge. Knowledge management will stimulate, in perspective, knowledge creation and knowledge acquisition from inside the organization as well as from the external environment. Exploration strategy is a key driving force for innovative firms. The focus strategy does not have a direct equivalent in the field of knowledge, but it approaches the *disruptive innovation strategies*. Although we started by considering the business generic strategies defined by Michael Porter (1985), we have to underline a big difference in the domain of intangible resources. Firms should not choose between exploitation and exploration strategies but should find a balance between them by developing a knowledge strategy ambidexterity (Raisch et al. 2009).

7.1.2 The Known-Unknown Matrix

The answer formulated by the former Secretary of Defense Donald Rumsfeld (2002) during the U.S. Department of Defense *News briefing* has become quite famous: "Reports that say that something hasn't happened are always interesting to me, because as we know, there are *known knowns*; there are things we know we know. We also know there are *known unknowns*; that is to say we know there are some things we do not know. But there are also *unknown unknowns*—the ones we don't know we don't know" (italics added). These expressions "known unknowns" and "unknown unknowns" generated a lot of critiques and debates from many journalists, writers, language experts, philosophers and people involved in economics and politics. However, these expressions reflect *the known-unknown paradox* that can be obtained by combining the level of awareness of what we know with the degree of known in the external world in a matrix (Dalkir 2005). Figure 7.1 presents an illustration of the known-unknown matrix with associated generic knowledge strategies.

The matrix incorporates four states of knowledgeable domains that can be phrased as follows:

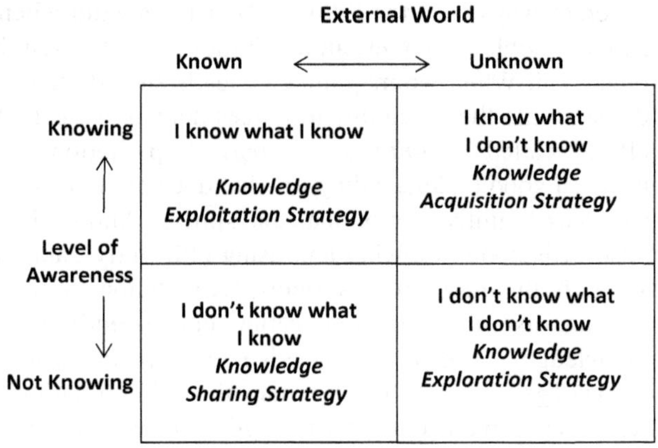

Fig. 7.1 The known-unknown matrix

- I know what I know.
- I know what I don't know.
- I don't know what I know.
- I don't know what I don't know.

The first two sentences reflect a static and finite world of knowledge, and a deterministic way of thinking (Bratianu 2007, 2015b). I know what I know because I am certain about my knowledge. Since certainty can be understood only in terms of conscious thinking, it results that I am considering my rational knowledge. The second sentence refers to the gap in my knowledge with respect to a given finite and static world of knowledge. I know what I don't know because I know how much I am supposed to know in this field, or in this life, about the external world. These first two sentences, which synthesize the "known" domains, substantially refer to explicit knowledge. The third domain is a little more difficult to comprehend since it considers both explicit and tacit knowledge. Since tacit knowledge reflects the experience we have it is hard to be aware of how much we know. I know that I can use the knowledge got through my experience but I don't know what exactly I know and how much I know. The fourth domain is about the infinity of knowledge in the external world and our practical impossibility to be aware of all of it. At the same time, it is about the absence of knowledge concerning the probable future events and phenomena.

For each domain we can associate some generic knowledge strategies each organization may develop in order to close the identified gaps between the known and the unknown fields. If we project some strategic objectives into the future and there is a strategic gap between where we are and where we want to be, then we should be able to define an associate knowledge gap which represents the "unknowns" from that matrix. Thus, for the first domain, we can label "known-knowns" the development of knowledge exploitation strategies which will enhance the efficiency of data, information and knowledge processing. For the second domain, we can label "known-unknowns" the case when we have to develop knowledge acquisition strategies which will help us in acquiring the knowledge we need in realizing the new products and services for achieving competitive advantage. The third domain can be labeled "unknown-knowns", and for it we need to implement knowledge sharing strategies through which tacit knowledge can be externalized and shared with others. The forth domain, that we may label "unknown-unknowns", requires knowledge creation which can be obtained through knowledge exploration strategies.

7.1.3 Organizational Knowledge Dynamics

The perspective of organizational knowledge dynamics has been initiated in Bratianu (2011), based on the metaphorical analysis of knowledge as energy, and then developed in Bratianu et al. (2011) by using the Analytic Hierarchy Process (AHP). *Organizational knowledge* is a semantic construct designed to reflect the integrated individual knowledge fields of all employees and the codified knowledge embedded in the procedures, routines, documents of intellectual properties, data bases and organizational culture (Becerra-Fernandez and Sabherwal 2010; Brown and Duguit 1998; Davenport and Prusak 2000; Nonaka and Takeuchi 1995; Spender 1996; Sveiby 2001). According to the multi-field theory (Bratianu 2015a), organizational knowledge is a result of the work performed by nonlinear integrators on the rational, emotional, and spiritual knowledge fields. The most powerful nonlinear integrators are leadership, management, and organizational culture. Management acts mostly on rational knowledge, leadership acts mostly on emotional knowledge, and organizational culture on spiritual knowledge. Knowledge is created at individual level and then through complex social processes it is amplified and structured at the team and organizational levels (Nonaka and Takeuchi 1995).

The knowledge-based theory of the firm (Nickerson and Zenger 2004; Spender 1996; Sveiby 2001; Tsoukas 1996) is conceived on the assumption that any organization can be viewed as a system of distributed knowledge, bounded by an interface which separates the internal fields of knowledge from the external fields of knowledge. This assumption is also shared by Schiuma (2009, p. 292), in his model of knowledge assets dynamics: "Every organization can be analyzed as a system made of knowledge elements, that is knowledge resources that are to some extent interdependent". Figure 7.2 shows such a holistic view of the firm which is open to knowledge transfer with respect to the business environment. From a strategic perspective, the firm must be in dynamic equilibrium with external fields of knowledge forces, and must have the capacity of responding fast to the rapid and unpredictable changes in the turbulent business environment. That means to achieve a positive variation of the level of organizational knowledge, where organizational knowledge results from the balance between inward fluxes of knowledge, knowledge creation and the outward fluxes of knowledge.

This model of organizational knowledge dynamics is similar to the conservation law of energy applied to an open system. However, there is an important difference

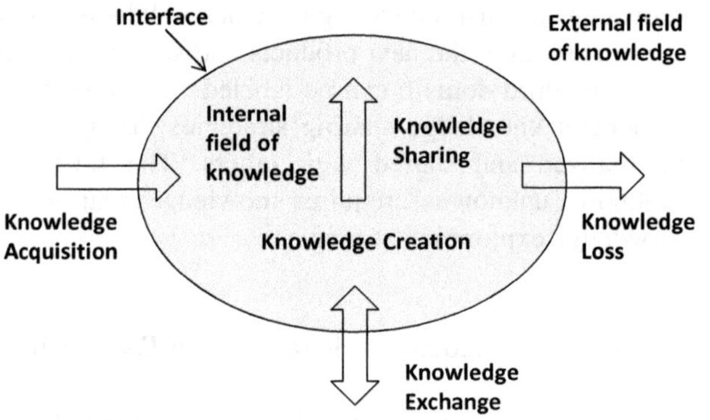

Fig. 7.2 Organizational knowledge dynamics

with respect to energy conservation since knowledge can be created and can be destroyed while energy can only be transformed from one form into another. The organizational knowledge dynamics evaluation is necessary in identifying the *knowledge gap* which is associated to the *strategic gap* with respect to strategic intention of the firm. As showed by Zack (1999, p. 135), "underlying a firm's strategic gap is a potential *knowledge gap*. That is, given a gap between what a firm must do to compete and what it can do, there may also be a gap between what the firm must know to execute its strategy and what it does know". Knowledge strategies are designed to close this strategic knowledge gap which is aligned with the business strategic gap. That "is essential for assuring that knowledge management efforts are being driven by and are supporting the firm's competitive strategy" (Zack 1999, p. 135). When intangible resources are dominant and the firm is knowledge intensive, knowledge strategy becomes the driving force of the company, as showed in the previous chapters.

7.2 Knowledge Exploitation Strategy

7.2.1 Organizational Ambidexterity

The exploitation strategy might be considered an oxymoron. On one hand, the concept of "exploitation" means to use efficiently the organizational resources for a short term gain, and on the other hand the concept of "strategy" means long term thinking. Since time is a continuum, the link between short term and long term thinking is done through a process of continuous adaptation by a systematic exploitation of the existing organizational knowledge. "Exploitation includes such things as refinement, choice, production, efficiency, selection, implementation, execution" (March 1991, p. 71). Knowledge exploitation strategy implies a continuous adaptation of the organization to the changeable business environment by making use efficiently of all intangible resources and capabilities existing in organization. Even though there are contributions to the knowledge field, they are incremental and contribute to the improvement of what has already been done without any dramatic impact on competitive advantage. Peter Senge (1999, p. 14) considers that knowledge exploitation can be seen as a "survival learning" process or "adaptive learning". It is an important process, but not sufficient for creating a competitive advantage: "*Adaptive learning* is important—indeed it is necessary. But for a learning organization, *adaptive learning* must be joined by *generative learning*, learning that enhances our capacity to create" (emphasis in original).

Generative learning can be realized by an exploration strategy, which means that an organization, aiming at achieving a competitive advantage, should create a balance between *knowledge exploitation* and *knowledge exploration* strategies. Some authors call this dynamic capability *organizational ambidexterity* (Andriopoulos and Lewis 2009; Cao et al. 2009; Raisch et al. 2009). "The third tension relates to *static* versus *dynamic* perspectives on ambidexterity. Although some research suggests that sequential attention should be paid to exploitation and exploration, the majority of organizational ambidexterity research presents a range of solutions that enables organizations to simultaneously pursue the two activities" (Raisch et al. 2009, p. 686). However, organizational ambidexterity is conditioned by the limited resources and capabilities which should be allocated for implementing these two complementary strategies and by the goal of shareholders of maximizing their profit for all investments made. Also, it is well-known that returns from knowledge exploitation are faster than those of knowledge exploration, and that they have a lower degree of uncertainty. What is good for shareholders in a short run might not be good for the whole organization in the long run. The strategic solution to this conflicting decision process is creating a dynamic equilibrium between knowledge exploitation and knowledge exploration since both of them are essential for the strategic positioning of organization (March 1991; Raisch et al. 2009). Moreover, results of knowledge exploitation can be fed up in knowledge exploration, and outcomes of knowledge exploration can be improved incrementally by knowledge exploitation.

As a strategy, knowledge exploitation can be implemented through a series of processes which have, as a common base, increasing efficiency of using the known organizational knowledge for improving the existing routines, procedures, products and services. It is the essence of the known-knowns domain of the awareness matrix presented in Fig. 7.1. In the following sections, we shall present two of the most successful applications of this strategy, namely knowledge codification and knowledge mapping.

7.2.2 Knowledge Codification

Knowledge codification implies transforming cognitive, emotional and spiritual knowledge into messages that can be understood by all employees of a certain organization. It occurs inside the organization but its consequences should be observed in both internal and external environment. Knowledge codification enables knowledge communication, knowledge use and reuse. Without data, information and knowledge codification communication between people, between people and computers, between computers and intelligent technological artifacts, or between organizations and their external business environment, would have not been possible. Codification is usually defined as a process of transforming an idea into an object, using a code that can range from an abstract to a metaphorical form. Starting from this assumption, knowledge codification is presented as a process that supports "the inscription of knowledge in symbolic forms" (Cacciatori et al. 2012). If natural language, numbers or analytical models are used as codes, then knowledge is converted into declarative statements, documents, databases, lessons learnt reports and best practices handbooks. If nonverbal language, images or graphical models are used as codes, knowledge is codified into videos, practices and conduits. The rich diversity of codes from natural to symbolic languages emphasizes the adaptable and contextual character of knowledge codification. However, the choice of codification models and technologies depends on the economic metrics and constrains. As emphasized by Cowan et al. (2000, p. 22), "In practice, the extent to which knowledge is codified is determined by incentives: the costs and benefits of doing so. For instance, many factors—such as, to take the simplest argument, the high cost of codifying a certain type of knowledge—can decrease the incentives to go further, by lowering the private rate of return on codification".

For some authors, codification is more focused on formal knowledge and its processing capability by a firm. For instance, Janicot and Mignon (2012, p. 6) define codification as "a process of storage, indexation and distribution of formal knowledge independently of any context. To complete this definition, the concept of codification can be broadened to include standardization of knowledge".

Knowledge codification constitutes an imperative for any organization which aims at achieving competitive advantage, especially for the knowledge intensive business service firms. It integrates both human and software agents with the dominant role played by people and not the technology. Knowledge codification enables verbal and nonverbal communication at individual and organizational levels. At individual level, verbal codification is used during *combination* processes while nonverbal codification appears during *socialization* processes as described by Nonaka and Takeuchi (1995) in the famous SECI model. Both verbal and nonverbal messages resulting from codification express encrypted ideas, experiences, beliefs and cultural values (Nonaka and Takeuchi 1995; Spender and Strong 2014; Stacey 2001).

Explicit knowledge is, in itself, a result of a mental codification process by using a natural language. Then, written knowledge is a result of codification by using grammatical codes and cultural routines. Organizational knowledge in its multiple forms is also a result of codification and integration of all employees' knowledge. It is a shared knowledge. Creating data, information and knowledge bases is possible only by using specific codes to structure, store and retrieve them: "rapid advancement in software agent technology has allowed for embedding organizational data, routines and processes into routines, repositories and function-creating opportunities for time-saving, duplication of effort, and consistency through rule-based reasoning" (Datta and Acar 2010, p. 48). All the rules and constraints are based on a specific logic which essentially represents a codifying process. However, coding data, information and knowledge implies decoding, which means the intervention of the human agents for their retrieval, reinterpretation and re-contextualization. In knowledge codification, human and information intelligent agents are synergistically bound as a result of systems thinking (Gharajedaghi 2006; Senge 1999) and nonlinear processes (Bratianu 2009; Gladwell 2000). Codification emerged as a necessity process for making possible communication in a social context, and as a means of increasing the cognitive value of organizational knowledge. However, that depends on the trust existing among people, among organization and its stakeholders in using codified knowledge. As remarked by Scarso and Bolisani (2012, p. 19), "the issue of trust is significant in KIBS-client interaction, especially when customized services delivered by means of a sparring relationship are involved". We shall discuss more about trust in a further section presenting the knowledge sharing strategy.

Considering the multi-field theory of knowledge (Bratianu 2015a) we have to consider not only rational or explicit knowledge but also emotional and spiritual knowledge. From this perspective, two domains of knowledge codification should be presented: dress codes and ethical codes. A dress code is a set of requirements an organization may formulate with respect to the way its employees should wear to work. Dress codes vary from organization to organization and range from formal to business casual, or casual. At one extreme a dress code might imply a uniform while at the other there are only some suggestions for casual and descent dress. A dress code incorporates rational, emotional and spiritual knowledge in different proportions. For instance, in a hospital, medical doctors usually have white colored dresses which mostly integrate rational and emotional knowledge, while in a university, professors and students wear, during special ceremonies, traditional caps and gowns which mostly incorporate emotional and spiritual knowledge. Dress codes are usually requested in firms where employees have interactions with customers, like law firms, accounting firms, fast food restaurants, shopping centers or airplanes. In workplaces where some employees interact with customers or clients, and others do not have these business interactions, organization may choose to have two different dress codes. IBM was famous not only for its computers but also for its organizational culture which requested a formal business attire: "It was well known throughout business circles that IBM salespeople—or, for that matter, any IBM employee—wore very formal business attire. Tom Watson established this rule when IBM was calling on corporate executives who—guess what—wore dark suits and white shirts!" (Gerstner 2003, p. 184). That dress code resisted up until 1995 when the new CEO Gerstner, Jr. abolished it.

IBM is also an excellent example of the way the founder Thomas Watson Sr. could impose the company his beliefs and values creating an organizational culture of respect, hard work, and ethical behavior. Reflecting on that aspect, Gerstner Jr. (2003, p. 182) remarked: "I came to see in my time at IBM, that culture isn't just one aspect of the game—it *is* the game. In the end, an organization is nothing more than the collective capacity of its people to create value. Vision, strategy, marketing, financial management—any management system, in fact—can set you on the right path and carry you for a while. But no enterprise—whether in business, government, education, health care, or *any* area of human endeavor—will succeed over the long haul if those elements aren't part of its DNA" (emphasized in original). Shared values create a framework for the decision making process which becomes a code of ethics for all employees. According to *Collins English Dictionary*, an ethical code is "a set of moral principles used to govern the conduct of a profession".

An *ethical code* for any organization represents a deep codification process which involves all types and forms of knowledge and is strongly interacting with its vision, mission, and strategizing process. For IBM the basic beliefs representing the backbone of its ethical code formulated by its founder and reinforced by Watson, Jr. are the following (Gerstner Jr. 2003, p. 184):

- Excellence in everything we do.
- Superior customer service.
- Respect for the individual.

A great organization is one with a great set of moral values which are shared by all its employees. It is valid not only for firms but also for not-for-profit organizations. A good example can be the world-class universities and their ethical codes. We shall illustrate this idea considering one of the most prestigious universities in the world—Harvard University. Since there are different formulations for different schools, we shall present the principles of the academic community of The Harvard Kennedy School posted on the official site of the university:

- Respect for all members of our community and for the space we share.
- Professionalism in all things, including the pursuit of intellectual and academic excellence.
- The recognition of the value of different opinions in our "free marketplace of ideas."
- Individual accountability for actions inconsistent with this *Code of Conduct*.
- Members of the community have a personal responsibility to integrate this code into all aspects of their experience.

In conclusion of this section, we would like to emphasize again that knowledge codification should not be limited only to explicit knowledge and information systems. Codification is a complex process which integrates in specific forms rational, emotional, and spiritual knowledge and contributes directly to the increase of organizational effectiveness as a dynamic capability able to produce and sustain the competitive advantage of the firm.

7.2.3 Knowledge Mapping

Knowledge exploitation as a strategy implies two essential conditions for knowledge management: a) to have a realistic evaluation of organizational knowledge and its distribution through the whole organization; b) to have the capability of using and re-using efficiently that available organizational knowledge. The first condition reflects the managerial capacity of evaluating the quantity, quality, distribution, and transferability of knowledge resources in all their form. That means to accept that knowledge is not uniform and homogeneous from a quality point of view and that it is compelling to distinguish between procedural knowledge and expertise, or between rational, emotional and spiritual knowledge. Although there are some methods proposed to measure knowledge directly (Bolisani and Oltramari 2012; Vallejo-Alonso et al. 2011), or indirectly by measuring the intellectual capital of an organization (Andriessen 2004; Guthrie et al. 2012; Sveiby 2010), the complexity of the problem delays the creation of a metric able to provide acceptable results (Bolisani 2016). We will discuss more about this topic in Chap. 8. Knowledge distribution is not uniform through an organization, since knowledge resides mostly in people and they have different levels of education and experience, and they are working in activity domains with different levels and types of knowledge. Transferability of knowledge depends on the form of knowledge and different individual and organizational barriers. While explicit knowledge can be transferred easily from one part of organization to another, tacit knowledge remains embodied in the employees' experience and can be shared through socialization or conversion to explicit knowledge (Bratianu 2015a; Nonaka and Takeuchi 1995).

The second condition of using and re-using the available organizational knowledge efficiently depends on the managerial intelligence and the tools available. *Knowledge mapping* is a complex process that integrates organizational intelligence, dynamic models and IT tools (Driessen et al. 2007; Kim et al. 2003; Lee and Fink 2013; Van den Berg and Popescu 2005). The result of this process is a *knowledge map* which becomes a navigating tool for organizational knowledge. "A knowledge map portrays the sources, flows, constraints, and sinks (losses or stopping points) of knowledge within an organization" (Liebowitz 2005, p. 77). The first generation of *knowledge maps* contains maps which are based on the metaphor of *knowledge as stock* representing only databases and sources of knowledge within the organization. These knowledge maps offer a *static* representation of knowledge distribution and related topics (documents, operational information, people) for the use of knowledge seekers and decision makers. From a very practical point of view, they are software programs based on a certain taxonomy and structure of organizational knowledge, and a specific logic of coding, storing and retrieving data, information and knowledge. Also, they can identify people with some profile (i.e. specific experience and expertise on given activity domains). Knowledge maps can have a hierarchical or neural structure. In the first case, the search is based on a top-down semantic search while, in the second case, the search is based on key words and neural-like branches. Instead of progressing linearly with the search from one level to another structure level, the neural maps provide semantic connections like the radiant thinking used in Mind Maps (Buzan and Buzan 1993, p. 59): "The Mind Map has four essential characteristics: a) the subject of attention is crystallized in a central image; b) the main themes of the subject *radiate* from the central image as branches; c) branches comprise a key image or key word printed on associated line, and topics of lesser importance are also represented as branches attached to higher branches; d) the branches form a connected nodal structure". The value of knowledge maps resides in the global image of the available (explicit and tacit) knowledge in organizations, and algorithms to find and use the needed knowledge. Thus, knowledge maps help to re-use critical knowledge and increase the exploitation efficiency.

The second generation of knowledge maps is based on the conceptual metaphor of *knowledge as stocks-and-flows*. A knowledge map of this category represents the sources of all types of knowledge with their main attributes, and the flows of knowledge from one part of organization toward another one. Kim et al. (2003, p. 36) define such a knowledge map as "a diagrammatic representation or corporate knowledge, having nodes as knowledge and links as the relationships between knowledge, and knowledge specification or profile". Thus, knowledge mapping evolved to include not only sources of information and knowledge but also the main knowledge flows within the organization. It is a mapping of the Nonakian dynamics of organizational knowledge (Nonaka 1994; Nonaka and Takeuchi 1995). Knowledge flows are generated and structured by business process flows. As Yoo et al. (2007, p. 107) remark, "Knowledge flows and business processes cannot be separated because knowledge is inputted and outputted through business processes. Knowledge flows inherit the feature and appearance of corresponding business processes".

A knowledge map consists of two components: a) diagram—graphical representation of knowledge, having nodes and linkages; b) specification—descriptive representation of knowledge. Knowledge maps help employees to search easily for the available knowledge within organization and use it such that they can optimize the cognitive work and increase productivity of knowledge processing. There are some clear advantages of building and using knowledge maps (Driessen et al. 2007; Kim et al. 2003; Lee and Fink 2013; O'Donnell et al. 2002):

- Codification and formalization of all knowledge inventories within an organization based on certain ontology and processing logic.
- Increasing knowledge retention from experts who retire, by making their expertise known and available to other knowledge seekers.
- Understanding of relationships between different knowledge sources or between knowledge sources and knowledge users. Also, easier understanding of cause and effect relationships.
- Efficient navigation across the organizational knowledge fields and reduction in cognitive load for knowledge seekers, due to their holistic representation of relationships among complex constructs.

- Stimulation of knowledge sharing and creation of communities of practice based on common interests. That results in an increase in the average level of organizational knowledge and in stimulated innovation.

In a synthetic manner, Kim et al. (2003, p. 37) conclude: "The knowledge map plays a key role in the KM project because it gives a knowledge profile (knowledge warehouse), knowledge link (navigation aids among knowledge), and expert finder".

7.3 Knowledge Acquisition Strategy

7.3.1 Knowledge Acquisition

Knowledge acquisition means finding ways of increasing the level of organizational knowledge by purchasing knowledge from the external business environment. In a strategic perspective this strategy contributes to closing the knowledge gap (Zack 1999) between what is available in the firm and what is needed for achieving a strategic objective. Knowledge acquisition is conceived as an alternative to knowledge creation, when the firm's human capital is not able to generate new knowledge or the process of knowledge creation is beyond the financial resources of the firm (Chaston and Mangles 2000; Davenport and Prusak 2000; Hoe and McShane 2010). Frequently, that situation happens when the firm is small and makes efforts to develop new products and services. Knowledge acquisition is a common strategy used by SMEs (Chan and Chao 2008; Desouza and Awazu 2006; Durst and Edvardson 2012). Knowledge acquisition in organizations spans a large spectrum of activities in concordance with the size of organization and its mission. It refers mostly to cognitive knowledge since emotional and spiritual knowledge is generated internally by people and their cultural values.

It represents just one activity of a broader and more complex process, as shown by Liao et al. (2010, p. 21): "Acquiring knowledge is the first activity in the broader activity of accepting knowledge from the external environment and transforming it into a representation that can be internalized, and/or used within the organization".

Knowledge acquisition is a frequently used strategy for entrepreneurial firms. As Studdard and Munchus (2009, p. 243) remark, "one of primary factors that hinder the formation and development of entrepreneurial firms is resource constrains. It is difficult for the entrepreneurial firm to sufficiently generate and acquire internal and external knowledge". Knowledge acquisition can be done in various ways, from buying access to scientific literature and knowledge bases, to hiring skilled people. For designing knowledge strategies, firms can purchase expertise from consulting companies and organize training programs for specific needs by hiring experts. For all firms (regardless their size), whose strategic intention is innovation, it is critical to acquire intellectual properties rights by purchasing patents. A successful strategy for small firms to increase knowledge acquisition with reasonable costs is to become part of some knowledge or *learning networks* (Chaston and Mangles 2000).

Network knowledge transfer focuses mostly on explicit rational knowledge. Emotional and spiritual knowledge has been used during the negotiation and formation of the learning network. Without a convergence of spiritual knowledge fields from member firms the construction and operation of the network is almost impossible. Spiritual knowledge is the driving force of the whole learning network, and emotional knowledge from each firm may become the catalyst of the learning process. It is interesting to note that learning processes are nonlinear, and that the network effect on knowledge transfer increases exponentially with the number of participants. Knowledge networks enlarge the boundary of the organizational knowledge fields and create the need of finding ways of integrating the internal and external knowledge. "With the knowledge integration model the skillful coordination, collaboration, and integration of vertical and horizontal boundaries promote the dynamic knowledge integration process, and build strategic innovation capability" (Kodama 2011, p. xii). An alternative of becoming a part of a business network is to create business alliances, which are forms of collaboration "where two or more organizations share resources and activities to pursue a strategy" (Johnson et al. 2011, p. 338). In such a business alliance, one of the strategic partners can share its knowledge. Thus, a business alliance has a critical role in the process of knowledge acquisition because it facilitates firm's access to best practices and shared markets (Grant 1996; Tsai 2001).

Knowledge acquisition depends on the *absorptive capacity* of the organization. The concept of absorptive capacity has been coined by Cohen and Levinthal (1990) and reflects the organization's capability to value, assimilate and apply knowledge coming from external sources. Whenever a firm has to establish the future business strategy or to elect the most efficient manner of adapting to market demand, it analyzes the business environment, identifies and captures the critical knowledge, and then develops the necessary structures and systems for using the acquired knowledge. Organizational absorptive capacity is based on individuals' absorptive capacity, and their integration derives from a management effort and from the use of IT systems that are employed in the firm. Based on its organizational absorptive capacity, a firm identifies the knowledge that must be acquired, determines the most efficient way to assimilate the selected knowledge and also balances the new knowledge with the previous one.

7.3.2 Knowledge Capturing

Knowledge capturing is a process through which trained people can extract valuable knowledge from experts and embed it in databases or intelligent software programs. Experts are individuals with a high level of knowledge and understanding in a given field of activity. By integrating all knowledge and experience, acquired by performing a series of intellectual activities, experts develop a significant expertise in time. This expertise is not proportional with the number of years spent in the same domain, because it is not a linear entity. Expertise is a nonlinear intangible which correlates with the variety of experiences and the quality of knowledge integrated through hard work and intelligent processing. Expertise is an excellent example of integration of rational, emotional, and spiritual knowledge, generated during a long period of time and stored as experience in our memory. "The bottom line truth is that tacit knowledge is not so much transferred as it is acquired and the process for acquiring tacit knowledge requires personal experience. There are no shortcuts. People don't become experts by reading what others have written, they become experts by doing" (Eucker 2007, p. 12).

Expertise is related to experience, and through experience to time, but not only quantitatively. Intensity of work and high motivation are also important factors in becoming an expert. Research performed in this field concluded that true expertise needs about ten thousands of working hours on the same topic or focused on the same type of activity (Levitin 2006). Discussing about expertise Gladwell (2008, p. 45) underlines the same idea for chess: "To become a chess grandmaster also seems to take about ten years. (Only the legendary Bobby Fischer got to that elite level in less than that amount of time: it took him nine years.) And what's ten years? Well, it's roughly how long it takes to put in ten thousands hours of hard practice. Ten thousands hours is the magic number of greatness".

Knowledge capturing from experts enables the realization of *expert systems* which are software tools to support decision making. "An expert system can provide people with advice by replacing part of the reasoning that is performed by experts. In fact, experts can use such a system themselves to reduce workload when there is too much to do and too little time" (Milton 2007, p. 3). In essence, an expert system is composed of a computable knowledge base of domain concepts and an interface engine of procedural rules *if-then*. Many expert system developers discovered that the difficulty is not to design and write a computer program but to elicit knowledge from experts. There was a "knowledge acquisition bottleneck" since knowledge engineers were able to create computer programs but they had no training in capturing knowledge from experts. To overcome their lack of experience in capturing knowledge from experts, knowledge engineers developed some automated knowledge acquisition systems called "shells". A shell is an interactive computer program that contains a series of questions aiming at extracting knowledge directly from experts. That means that each shell can be used only for a well-defined class of problems for which there is a significant knowledge captured. Moreover, these expert systems deal only with rational knowledge although any decision process is conditioned by the integration of rational, emotional and spiritual knowledge (Bratianu 2015a). Research performed in cognitive sciences concerning knowledge capturing converged to a new activity called *Cognitive Task Analysis* (Clark et al. 2008; Hoffman and Lintern 2006). Although there are several methods of performing Cognitive Task Analysis, the main stages of the process are essentially the same: a) search for and collect preliminary knowledge; b) identify knowledge representation; c) use a combination of knowledge elicitation methods; d) analyze, verify and interpret data acquired; e) put the final result into an adequate format. The outcome of a knowledge elicitation process is a knowledge base, a knowledge store, a knowledge repository or an ontology. From Knowledge

Management point of view the main limitation of any expert system is the emphasis on rational knowledge and on the *if-then* logic.

7.3.3 Knowledge Retention

Knowledge retention is the result of applying knowledge captured from experienced workers who retire. As shown in Fig. 7.2, organizational knowledge should find a balance by compensating knowledge loss (due to people who leave the firm because they retire or are fired, for example during an economic crisis). Especially when thousands of employees suddenly leave a company, the dynamic equilibrium of organizational knowledge is severely destroyed. A famous example comes from Boeing (DeLong 2004, p. 19): "After Boeing offered early retirement to 9000 senior employees during a business downturn, an unexpected rush of new commercial airplane orders left the company critically short of skilled production workers. The knowledge lost from veteran employees combined with the inexperience of their replacements threw the firm's 737 and 747 assembly lines into chaos". Another example is given by DeLong (2004) and concerns the impact of knowledge loss on organizational knowledge dynamics of Delta Airlines. The company decided, during a downsizing operation in 2001, to offer attractive packages to senior workers for leaving the company. As a result, 11,000 employees voluntarily left the company, including about 1200 aviation maintenance technicians, many with 20–30 years of experience at Delta Airlines. Although the business of the company improved as a result of cost cutting with salaries, the company later suffered from a severe knowledge loss, with negative long term consequences.

Knowledge retention is a complex process through which organizations can reduce knowledge loss. DeLong (2004) analyzes knowledge retention strategies and group them into four main categories: a) human resources, processes and practices; b) knowledge transfer practices; c) knowledge recovery initiatives; and d) information technology applications to capture, store and share knowledge. In the first group we have all the methods and mechanisms that human resource departments may use to create a long-term approach for reducing the individual and organizational knowledge loss qualitatively and quantitatively. That means to improve the system for evaluating employees' competences and to analyze where the organization is at risk with losing critical knowledge. That analysis should be complemented with an extensive career development and managerial succession planning processes. Another issue related to knowledge retention is the succession of departing leaders. Intelligent organizations create plans for leadership succession to make a slowly and efficient transfer of knowledge from the leaders who retire towards the new generation of leaders. That means to develop a process of intergenerational learning and to create a necessary culture for making it efficient (Lefter et al. 2011; Tichy 1997). Great leaders like Jack Welch of General Electric, Andy Grooves of Intel, and Roger Enrico of PepsiCo spent a lot of their agenda's time in teaching younger generations of leaders. As Tichy recalls (1997, p. 56): "Jack Welch of General Electric is one of the most dedicated teachers I know. For twenty years he has made biweekly visits to GE's Crotonville executive training center to enter into dialogue with thousands of his employees each year. His schedule was also filled with hundreds of video conferences, meetings, factory visits and workshop sessions". Also, Jack Welch designed a comprehensive program to prepare the next CEO of GE when he will retire, and Jeff Immelt is the proven success of this process.

7.4 Knowledge Sharing Strategy

7.4.1 Knowledge Sharing

Knowledge sharing is a strategy that increases the average level of organizational knowledge and contributes directly to the increase of organizational entropy. Intelligent and creative organizations discovered that achieving competitive advantage through innovation needs a higher level of knowledge and a higher value for their organizational entropy. As a process, knowledge sharing contributes to organizational knowledge creation from individuals' knowledge. "Organization cannot create knowledge on its own without the initiative of the individual and the interaction that takes place within the group. Knowledge can be amplified or crystallized at the group level through dialogue, discussion, experience sharing, and observation" (Nonaka and Takeuchi 1995, p. 13). Knowledge sharing is a process by which an individual is willing to share his or her experience with others without expecting any financial reward out of it. It is not an imposed activity by the managers like a working task, although it could be stimulated by creating an organizational culture favorable to it. Knowledge sharing involves activities of transforming or disseminating knowledge from one person to another, to a group of people, or to a whole organization. According to Cyr and Choo (2010, p. 825), knowledge sharing in organizations may be viewed "as the behavior by which an individual voluntarily provides other members of the organization with access to his or her knowledge and experiences. Knowledge sharing encompasses a broad range of behaviors that are complex and multi-faceted". Its importance comes from the fact that knowledge sharing links the individual knowledge fields where knowledge is generated, to the organizational level where knowledge is applied and attains value.

The opposite attitude to knowledge sharing is *knowledge hoarding* (Cyr and Choo 2010) which reflects egoism, lack of trust in other people and fear of losing power. Knowledge sharing is a voluntarily process but it depends on many personal and organizational factors that may stimulate it or may inhibit it. Szulanski (1995, 2000) extended the meaning of the concept of *stickiness* introduced by von Hippel (1994) for information transfer within an organization to knowledge sharing and transfer. "The assessment of the degree of difficulty experienced in a transfer is likely to reflect the number and intensity of those distinct moments of difficulty. Other things equal, a transfer is more likely to be perceived as difficult or sticky when efforts to resolve transfer problems become noteworthy" (Szulansky 2000,

p. 11). Knowledge stickiness appears especially in organizations where there is a culture of fierce individual competition and a fear of losing a certain usefulness if one's expertise is shared with a group of people. At the limit, the person who shared his or her experience is not of interest anymore and he or she can be fired at any time. In organizations where there is a team culture and cooperation is valued both by managers and employees, knowledge sharing is a current practice. In such situations, there is a culture of trust and of rewarding for people who share their knowledge. Nonaka and Takeuchi (1995) show that much of the business success of Japanese companies resides in their organizational culture of team work and knowledge sharing.

Trust is a powerful concept that has been used in many fields of activity, and defined in different ways. For instance, economists define trust in terms of quantitative aspects which can be measured using economic metrics, while psychologists define trust in terms of qualitative attributes of trustors and trustees. Sociologists focus on the quality of relationship between people and on the social context that influences them. One of the classic definitions which got some popularity has been formulated by Gambetta (Castelfranchi and Falcone 2010, p. 19): "Trust is the subjective probability by which and individual, A, expects that another individual, B, performs a given action on which its welfare depends". The definition focuses on the subjective probability of an individual which cannot be computed mathematically and reduces in practice to a personal belief. A more developed model of trust has been formulated by Mayer et al. (1995). It switches the focus from the attributes of trustor to the quality of the relationship between trustor and trustee and introduces the aspect of vulnerability. In their view, trust is "the willingness of a party to be vulnerable to the actions of another party based on the expectation that the other will perform a particular action important to the trustor, irrespective of the ability to monitor or control that other party" (Mayer et al. 1995, p. 712). It is evident that trust is a complex concept which integrates attributes of both trustor and trustee as well as the quality of their relationship in a given social context. Since the trustor becomes vulnerable to the possible negative consequences from the trustee, we can now understand why, in an organizational culture with a low level of trust, people are not willing to share their knowledge. Education, training and solid rewording systems should be used for a long time in organizations to build the necessary climate of trust and to stimulate knowledge sharing.

7.4.2 Communities of Practice

Communities of practice are not new ideas. They have been always present in the human history under diverse forms and structures. For instance, during the Middle Age, there were craft guilds that played similar roles like today's professional communities of practice. They disappeared as a result of the industrial revolution, but communities of practice continued to develop in almost any aspect of human life. According to Wenger et al. (2002, p. 4), "Communities of practice are groups of people who shape a concern, a set of problems, or a passion about a topic, and who deepen their knowledge and expertise in this area by interacting on ongoing basis". People who create these communities of practice can work together in the same company or not. Their institutional affiliation is not important. They find value in being together and sharing their rational, emotional, and spiritual knowledge as well as their aspirations to achieve some strategic objectives. Emotional and spiritual knowledge have also an important role in creating the social gravity field since there are no compulsory forces to act upon those who join a community of practice. The high level of trust and the common cognitive interests stimulate knowledge sharing and knowledge creation. Traditionally, communities of practice have often been groups of people who share face-to-face their knowledge. More recently, communities of practices also tend to become virtual networks where knowledge is shared by using all opportunities offered by advanced information systems and technologies (North and Gueldenberg 2011; O'Dell and Hubert 2011; Pasher and Ronen 2011).

Knowledge sharing in communities of practice stimulates learning and knowledge creation. As a result, these communities of practice may become knowledge or learning communities. North and Gueldenberg (2011, p. 149) define a knowledge community as "a group of people existing over a relatively long period who have interest in a common domain and want to develop and share knowledge together. Participation is voluntary and personal. Knowledge communities are formed around specific topics". Communities of practice may have short or long life cycles, they may become a successful project or a failure. Like any social construct, a successful community of practice needs to satisfy some requirements (North and Gueldenberg 2011):

- A well-defined *domain of knowledge sharing* that is attractive to a large spectrum of people.
- *A leader* able to create an attraction field of interests around him and a high level of social trust.
- *A critical mass* of people gathering and sharing knowledge.
- *An agenda of events* which can be improved continuously.
- *A rewording system* such that the most active participants to feel that their efforts are appreciated by the other members.
- *A website, newsletter or other publications.* These are essential for creating a dynamic communication between members of community.

The leader plays an essential role in designing that community of practice and attracting people as a result of his or her recognized expertise in that knowledge sharing domain. The leader should be able to create the necessary critical mass of participants and an interesting agenda of events able to keep alive the community. In some online communities of practice, created by using some smart websites, the leaders do not appear directly, but their expertise has been embedded in the structure and functionality of the online platforms.

It is interesting to see how great companies designed and implemented knowledge sharing models and developed communities of practice. We shall consider for illustration the case of ConocoPhillips (O'Dell and Hubert 2011) which is an international, integrated energy company and the third-largest oil and gas company in the United States. The company has about 30,000 employees working in over 30 countries. "Knowledge sharing methods were adapted to promote functional excellence and leverage knowledge across organization. Knowledge sharing sponsorship is now organization-wide and supported by all business streams" (O'Dell and Huber 2011, p. 163). The company developed a culture able to support and stimulate knowledge sharing as a means of learning for all employees. The company created more than 120 communities of practice known as networks of excellence. These communities are aligned in their activities with the business processes contributing to achieving competitive advantage. The experience obtained so far in running these networks of excellence reveal the following key success factors (O'Dell and Hubert 2011, p. 166):

1. Leadership and sponsorship.
2. A clear business case with a well-defined knowledge domain.
3. Adequate resources and defined roles.
4. Member engagement.
5. Deliverables and activities.
6. The development of trusted relationships.
7. Knowledge transfer processes.
8. Supporting information technology.
9. A system for motivation, recognition of results, and rewards for performance.
10. Network measurements for keeping track of each member activity.

The whole knowledge sharing system and all 120-plus networks of excellence are managed by a team of six experts. "The team addresses strategic goals and leverages resources across the organization. Working with all business streams and functional units, it is responsible for maintaining established networks and managing associated training, metrics, and portal sites" (O'Dell and Hubert 2011, p. 167). However, by far the most important result of the company vision in implementing that knowledge sharing system is the development of a knowledge sharing culture, based on mutual trust. To build trust, the company organizes face-to-face meetings and networking where people get the feeling of being members of the same community. However, there is the challenge of people working in different geographical zones with different cultural values and time zones. For instance, employees from western countries feel comfortable with posting questions and comments, while employees from Asia-Pacific region don't feel that way since they have a team culture and find real difficult to single them out by asking questions. Leaders of each community can try to find out solutions to overcome these cultural barriers.

7.5 Knowledge Exploration Strategy

7.5.1 Knowledge Creation

Although people enjoy living and working in a comfortable zone of known-knowns, the new turbulent business landscape increasingly imposes to search for the unknown-unknowns zone, which features a high levels of uncertainty and risks. Emergent strategies replace the deliberate ones and knowledge exploration strategies replace the exploitation strategies. "The essence of exploration is experimentation with new alternatives. Its returns are uncertain, distant, and often negative" (March 1991, p. 85). Exploration means searching for new knowledge and ways of increasing the level of organizational knowledge. Knowledge management will stimulate knowledge creation and knowledge acquisition in perspective, from inside the organization as well as from external environment. Open innovation is already a well-established process of acquiring new knowledge from external contributors. However, searching for new knowledge is a costly strategy and the outcomes are not certain. The risks associated to knowledge exploration are significantly higher than those associated with knowledge exploitation of organizational knowledge. Top management should be guided by a strong vision concerning the exploration of new ventures to produce goods and services. Knowledge exploration became a key strategy for innovative companies. Steve Jobs was such a visionary leader who wanted to change the world. "Under Job's leadership, Apple has earned a reputation as one of the most innovative companies in technology. Business Week in 2007 named Apple the most innovative company in the world, beating Google, Toyota, Sony, Nokia, Genentech, and a host of other A-list companies" (Kahney 2008, p. 179).

The most important contribution of an exploration strategy is given by knowledge creation. According to the Data-Information-Knowledge-Wisdom (DIKW) model, information is a result of processed data and knowledge an outcome of the processed information (Davenport and Prusak 2000; Jashapara 2011; Rowley 2007). In this conceptual framework, knowledge creation means information processing. However, the domain of information processing belongs to information science where *information* is the pivotal concept and it is defined based on Shannon mathematical theory of communication (Bratianu 2015a). Since there are different perspectives in information science and in knowledge management concerning the meaning of *information*, we shall confine our discussion to the knowledge management approach, and we shall present the main ideas of famous Nonaka's theory of knowledge creation dynamics (Nonaka 1994; Nonaka and Takeuchi 1995; Nonaka et al. 2008). The model is based on a series of knowledge transformations which can be represented on a diagram defined by epistemological and ontological dimensions. The epistemological dimension reflects the individual contribution to knowledge creation, while the ontological dimension reflects the social contribution. This way, Nonaka creates a synthesis between the psychological and sociological perspectives of knowledge creation.

The core of the Nonaka's theory of knowledge creation is the SECI model, which is composed of four conversion processes of tacit and explicit knowledge (Nonaka 1994). When people share common goals, they can form communities of practice, or communities of business processes, which contribute to the amplification and development of new knowledge. These communities define the ontological dimension of the model. Now, considering this epistemological-ontological knowledge space, "a spiral model of knowledge creation is proposed which shows the relationship between the epistemological and ontological dimensions of knowledge creation. This spiral illustrates the creation of a new concept in terms of a continual dialogue between tacit and explicit knowledge" (Nonaka 1994, p. 15).

Nonaka's model shows the important role played by social interactions and by the organizational framework of communication between employees. That becomes critical in knowledge intensive business processes (KIBP) where knowledge fluxes could attain high levels of intensity. "Within KIBP, it is the human ability to interpret the information obtained and transform the information to knowledge, thus providing the individual with the opportunity to further develop their own intuition and innovation based on KIBP experiences" (Little and Deokar 2016, p. 861). From a psychological perspective, this is a learning process integrating the past experience and knowledge with their future expectations and the business needs (Salmador and Florin 2013). Individual learning transforms through social interaction into a social learning process which amplified up to the organizational level becomes organizational learning (Argote 2013; Argyris 1999; Crossan et al. 1999).

7.5.2 Knowledge Co-creation

Knowledge co-creation emerged as a new paradigm for understanding the new cooperation processes between firms and their stakeholders. From a process of knowledge creation centered on the firm's R&D capability and embedding that knowledge in new products and services, we face today a transition toward a process of knowledge creation by the firm in partnership with its stakeholders. "Co-creation is the process where more than one party systematically joins forces to interact, learn and share information to create value" (Kennedy and Guzman 2016, p. 313). These co-creation phenomena have changed the way business strategies are designed and implemented (Kao et al. 2016; Millspaugh and Kent 2016; Paswan et al. 2014; Ramaswamy and Ozcan 2013; Verleye 2015). As a consequence, the process of value creation is not centered on the firm anymore but on its working relationships with customers and other stakeholders involved in the chain production. Consumers want to be involved in a series of activities related to product design, production and marketing, activities done so far only by firms. As Ramaswamy and Ozcan (2013, p. 7) remark, "In more and more firms, strategy making has become a joint process of co-creative discovery, as enterprises devise and develop new opportunities together with customers, partners and other stakeholders".

The knowledge process now extends beyond the boundary of the firm and integrates knowledge from the external business environment in new and attractive ways for all participants. The co-creation phenomena developed in an accelerated way especially in the service field, where service dominant logic stimulates knowledge exchange and final consumers enjoy having their expectations better fulfilled by the firms. This emerging consumer empowerment is mediated efficiently by social media networks (Kao et al. 2016; Kennedy and Guzman 2016). The co-creation phenomenon increases the firm's entropy very much, by reducing its full control on the production process. That means that there is a need for a new type of management able to give away some power to the participants in the co-creation process but to keep the overall control on the chain of designing, production and marketing activities. Kao et al. (2016) identify, in a generic co-creation process, five significant stages: interact—creating attractive conditions for interaction; engage—building user trust, loyalty and a sense of belonging; propose—enhancing knowledge sharing and users contribution; act—developing consensus and the participation to collective innovation; realize—evaluation of the result of the co-creation process.

Knowledge exploration strategy becomes essential in the turbulent business environment since the sustainable competitive advantage cannot be achieved by using the old success business formulas. New visions and explorations are necessary but not only within the internal business environment; explorations should extend to the external business environment where stakeholders can play an important role in knowledge co-creation and value co-creation. Closed innovation should be replaced by open innovation and customers should be part of that new and rewarding process. That means a new type of leadership and strategizing able to deal with a higher level of entropy and uncertainty.

Approaches to knowledge co-creation can be different, in relation to the characteristics of product or services, to the attitude of companies, and to their position in the value chain. Recent studies (Paiola et al. 2013) have collected evidence of the different possible approaches of companies aiming at acquiring and exchanging knowledge with external business partners, including suppliers and customers.

7.6 Conclusion

Knowledge acquisition strategy is useful for closing the knowledge gap between what a firm knows and what it is needed to be known for achieving competitive advantage. Knowledge acquisition comes as a first choice when the organization does not have a critical mass for knowledge creation or closing the knowledge gap requests both knowledge acquisition and knowledge creation. Knowledge acquisition implies purchasing knowledge from the external business environment by using different methods and practices. One of the most efficient methods is creating a business network with other firms or becoming a part of such a network. A network is an enabler for creating knowledge fluxes and an efficient balance between those which cross the organization interface in both directions (i.e. inward and outward). Knowledge acquisition is an attractive strategy for SMEs and especially for those which are entrepreneurial and innovative. The level of acquired knowledge depends on the absorptive capacity of each organization, which integrates both human and technology factors. Much of the organizational knowledge is stored within individuals who use it performing their tasks and playing the competition roles. Experience and expertise of many people remain a valuable intellectual capital potential without an efficient contribution to the new products and services. The only solution to making all that knowledge available throughout the organization is building a culture of trust and stimulating knowledge sharing. That can be achieved by encouraging people to participate in communities of practice where they can share their experience with others and learn new knowledge from them at the same time. Knowledge sharing embraces all forms of knowledge (i.e. rational, emotional, and spiritual knowledge). Finally, exploration strategy comes with knowledge creation, so that firms can sustain their competitive advantage in a turbulent business environment. Moreover, in designing new products and services, firms open themselves toward their customers and other stakeholders, to work together and co-create them. Open innovation is replacing the old system of closed innovation, and through co-creation, firms can use the potential knowledge residing in the external business environment.

References

Andriessen, D. (2004). *Making sense of intellectual capital: Designing a method for valuation of intangibles.* Amsterdam: Elsevier.

Andriopoulos, C., & Lewis, M. (2009). Exploitation-exploration tensions and organizational ambidexterity: Managing paradoxes of innovation. *Organization Science, 20*(4), 696–717.

Argote, L. (2013). *Organizational learning: Creating, retaining and transferring knowledge* (2nd ed.). New York: Springer.

Argyris, C. (1999). *On organizational learning* (2nd ed.). Oxford: Blackwell Business.

Becerra-Fernandez, I., & Sabherval, R. (2010). *Knowledge management: Systems and processes.* New York: ME Sharpe.

Bolisani, E. (2016, September 1–2). Methods of measuring knowledge: Analysis and classification. In S. Moffett & B. Galbraith (Eds.), Proceedings of the 17th European Conference on Knowledge Management, Ulster University, Northern Ireland, UK (pp. 91–100).

Bolisani, E., & Oltramari, A. (2012). Knowledge as a measurable object in business context: A stock-and-flow approach. *Knowledge Management Research and Practice, 10*(3), 275–286.

Bratianu, C. (2007). Thinking patterns and knowledge dynamics. In B. Martins & D. Remenyi (Eds.), *Proceedings of the 8th European Conference on Knowledge Management* (Vol. 1, pp. 152–157). Reading, MA: Academic Conferences Publishing International.

Bratianu C (2009, April 28–29) The frontier of linearity in the IC metaphor. In C Stam, D Andriessen (eds) Proceedings of the European Conference on Intellectual Capital, Inholland University of Applied Sciences, Haarlem, The Netherlands, pp. 97-103. Academic Publishing Limited, Reading, MA.

Bratianu, C. (2011). Changing paradigm for knowledge metaphors from dynamics to thermodynamics. *Systems Research and Behavioral Science, 28*(2), 160–169.

Bratianu, C. (2015a). *Organizational knowledge dynamics: Managing knowledge creation, acquisition, sharing, and transformation.* Hershey, PA: IGI Global.

Bratianu, C. (2015b). Developing strategic thinking in business education. *Management Dynamics in the Knowledge Economy, 3*(3), 409–429.

Bratianu, C., Agapie, A., & Orzea, I. (2011). Modeling organizational knowledge dynamics: Using analytic hierarchy process (AHP). *Electronic Journal of Knowledge Management, 9*(3), 236–247.

Brown, J. S., & Duguit, P. (1998). Organizing knowledge. *California Management Review, 40*(3), 90–110.

Buzan, T., & Buzan, B. (1993). *The Mind Map book: How to use radiant thinking to maximize your brain's untapped potential*. London: Penguin Group.

Cacciatori, E., Tamoschus, D., & Grabher, G. (2012). Knowledge transfer across projects: Codification in creative, high-tech and engineering industries. *Management Learning, 43*(3), 309–331.

Cao, Q., Gedajlovic, E. R., & Zhang, H. (2009). Unpacking organizational ambidexterity: Dimensions, contingencies, and synergistic effects. *Organization Science, 20*(4), 781–796.

Castelfranchi, C., & Falcone, R. (2010). *Trust theory: A socio-cognitive and computational model*. Chichester: Wiley.

Chan, I., & Chao, C. K. (2008). Knowledge management in small and medium-sized enterprises. *Communications of the ACM, 51*(4), 83–88.

Chaston, I., & Mangles, T. (2000). Business networks: Assisting knowledge management and competence acquisition within UK manufacturing firms. *Journal of Small Business and Enterprise Development, 7*(2), 160–170.

Clark, R. E., Feldon, D. F., van Merriënboer, J. J. G., Yates, K. A., & Early, S. (2008). Cognitive task analysis. In J. M. Spector & M. D. Driscall (Eds.), *Handbook of research on educational communications and technology* (3rd ed.). New York: Taylor & Francis.

Cohen, W. M., & Levinthal, D. A. (1990). Absorptive capacity: A new perspective on learning and innovation. *Administrative Science Quarterly, 35*(1), 128–152.

Cowan, R., David, P. A., & Foray, D. (2000). The explicit economics of knowledge codification and tacitness. *Industrial and Corporate Change, 9*(2), 211–253.

Crossan, M. M., Lane, H. W., & White, R. E. (1999). An organizational learning framework: From intuition to institution. *The Academy of Management Review, 24*(3), 532–537.

Cyr, S., & Choo, C. W. (2010). The individual and social dynamics of knowledge sharing: An exploratory study. *Journal of Documentation, 66*(6), 824–846.

Dalkir, K. (2005). *Knowledge management in theory and practice*. New York: Elsevier.

Datta, P., & Acar, W. (2010). Software and human agents in knowledge codification. *Knowledge Management Research and Practice, 8*(1), 45–60.

Davenport, T. H., & Prusak, L. (2000). *Working knowledge: How organizations manage what they know*. Boston: Harvard Business School Press.

DeLong, D. W. (2004). *Lost knowledge: Confronting the threat of an aging workforce*. Oxford: Oxford University Press.

Desouza, K. C., & Awazu, Y. (2006). Knowledge management at SMEs: Five peculiarities. *Journal of Knowledge Management, 10*(1), 32–43.

Driessen, S., Huijsen, W. O., & Grootveld, M. (2007). A framework for evaluating knowledge-mapping tools. *Journal of Knowledge Management, 11*(2), 109–117.

Durst, S., & Edvardson, I. R. (2012). Knowledge management in SMEs: A literature review. *Journal of Knowledge Management, 16*(6), 879–903.

Eucker, T. R. (2007). Understanding the impact of tacit knowledge loss. *KM Review, 10*(2), 10–13.

Gerstner Jr., L. V. (2003). *Who says elephants can't dance? How I turned around IBM*. London: Harper Collins.

Gharajedaghi, J. (2006). *Systems thinking: Managing chaos and complexity*. New York: Elsevier.

Gladwell, M. (2000). *The tipping point: How little things can make a big difference*. London: Little Brown Book.

Gladwell, M. (2008). *Outliers: The story of success*. London: Penguin Books.

Grant, R. (1996). Toward a knowledge based theory of the firm. *Strategic Management Journal, 17*(Special Issue), 109–122.

Guthrie, J., Ricceri, F., & Dumay, J. (2012). Reflections and projections: A decade of intellectual capital accounting research. *The British Accounting Review, 44*(2), 68–82.

Hoe, S. L., & McShane, S. (2010). Structural and informal knowledge acquisition and dissemination in organizational learning. An exploratory analysis. *The Learning Organization, 17*(4), 364–386.

Hoffman, R. R., & Lintern, G. (2006). Eliciting and representing the knowledge of experts. In K. A. Ericsson, N. Charness, P. Feltovich, & R. R. Hoffman (Eds.), *Cambridge handbook of expertise and expert performance* (pp. 203–222). New York: Cambridge University Press.

Janicot, C., & Mignon, S. (2012). Knowledge codification in audit and consulting firms: A conceptual and empirical approach. *Knowledge Management Research and Practice, 10*(1), 4–15.

Jashapara, A. (2011). *Knowledge management: An integrated approach* (2nd ed.). London: Financial Times/Prentice Hall.

Johnson, G., Whittington, R., & Scholes, K. (2011). *Exploring strategy: Text & cases* (9th ed.). London: Financial Times/Prentice Hall.

Kahney, L. (2008). *Inside Steve's brain*. London: Atlantic Books.

Kao, T. Y., Yang, M. H., Wu, J. T. B., & Cheng, Y. Y. (2016). Co-creating value with consumers through social media. *Journal of Services Marketing, 30*(2), 141–151.

Kennedy, E., & Guzman, F. (2016). Co-creation of brand identities: Consumer and industry influence and motivations. *Journal of Consumer Marketing, 33*(5), 313–323.

Kim, S., Suh, E., & Hwang, H. (2003). Building the knowledge map: An industrial case study. *Journal of Knowledge Management, 7*(2), 34–45.

Kodama, M. (2011). *Knowledge integration dynamics: Developing strategic innovation capability*. Singapore: World Scientific.

Lee, J., & Fink, D. (2013). Knowledge mapping: Encouragements and impediments to adoption. *Journal of Knowledge Management, 17*(1), 16–28.

Lefter, V., Bratianu, C., Agapie, A., Agoston, S., & Orzea, I. (2011). Intergenerational knowledge transfer in the academic environment of knowledge-based economy. *Amfiteatru Economic, 13* (30), 392–403.

Levitin, D. J. (2006). *This is your brain on music: The science of human obsession*. New York: Dutton.

Liao, S. H., CC, W., DC, H., & Tsui, K. A. (2010). Relationships between knowledge acquisition, absorptive capacity and innovation capability: An empirical study on Taiwan's financial and manufacturing industries. *Journal of Information Science, 36*(1), 19–35.

Liebowitz, J. (2005). Linking social network analysis with the analytic hierarchy process for knowledge mapping in organizations. *Journal of Knowledge Management, 9*(1), 76–86.

Little, T. A., & Deokar, A. V. (2016). Understanding knowledge creation in the context of knowledge-intensive business processes. *Journal of Knowledge Management, 25*(5), 858–879.

March, J. G. (1991). Exploration and exploitation in organizational learning. *Organization Science, 2*(1), 71–87.

Mayer, R. C., Davis, J. H., & Schoorman, F. D. (1995). An integrative model of organizational trust. *The Academy of Management Review, 20*(3), 709–734.

Millspaugh, J., & Kent, A. (2016). Co-creation and the development of SME designer fashion enterprises. *Journal of Fashion Marketing and Management, 20*(3), 322–338.

Milton, N. R. (2007). *Knowledge acquisition in practice: A step-by-step guide*. Berlin: Springer.

Nickerson, J. A., & Zenger, T. R. (2004). A knowledge-based theory of the firm: The problem-solving perspective. *Organization Science, 15*, 617–632.

Nonaka, I. (1994). A dynamic theory of organizational knowledge creation. *Organization Science, 5*(1), 14–37.

Nonaka, I., & Takeuchi, H. (1995). *The knowledge-creating company: How Japanese companies create the dynamics of innovation*. Oxford: Oxford University Press.

Nonaka, I., Toyama, R., & Hirata, T. (2008). *Managing flow: A process theory of the knowledge-based firm*. Houndmills: Palgrave Macmillan.

North, K., & Gueldenberg, S. (2011). *Effective knowledge work: Answers to the management challenges of the 21st century*. Bingley: Emerald Group Publishing.

O'Dell, C., & Hubert, C. (2011). *The new edge in knowledge: How knowledge management is changing the way we do business*. Hoboken, NJ: Wiley.

O'Donnell, A. M., Dansereau, D. F., & Hall, R. H. (2002). Knowledge maps as scaffolds for cognitive processing. *Educational Psychology Review, 14*(1), 71–86.

Paiola, M., Bolisani, E., & Scarso, E. (2013). Characterisation of knowledge-based networking strategies: Insights from the KIBS sector. *International Journal of Networking and Virtual Organisations, 13*(3), 263–281.

Pasher, E., & Ronen, T. (2011). *The complete guide to knowledge management: A strategic plan to leverage your company's intellectual capital*. Hoboken, NJ: Wiley.

Paswan, A. K., D'Souza, D., & Rajamma, R. K. (2014). Value co-creation through knowledge exchange in franchising. *Journal of Services Marketing, 28*(2), 116–125.

Porter, M. (1985). *Competitive advantage: Creating and sustaining superior performance*. New York: The Free Press.

Porter, M. (1996). What is strategy? *Harvard Business Review, 74*(November–December), 61–78.

Raisch, S., Birkinshaw, J., Probst, G., & Tushman, M. L. (2009). Organizational ambidexterity: Balancing exploitation and exploration for sustained performance. *Organization Science, 20*(4), 685–695.

Ramaswamy, V., & Ozcan, K. (2013). Strategy and co-creation thinking. *Strategy & Leadership, 41*(6), 5–10.

Rowley, J. (2007). The wisdom hierarchy: Representations of DIKW hierarchy. *Journal of Information Science, 33*(2), 163–180.

Rumsfeld, D. (2002). *Press conference at the U.S. Department of Defense*. Accessed January 10, 2014, from http://www.defense.gov/transcripts/transcript.aspx?trasncriptid=2636

Salmador, M. P., & Florin, J. (2013). Knowledge creation and competitive advantage in turbulent environments: A process model of organizational learning. *Knowledge Management Research and Practice, 11*(4), 374–388.

Scarso, E., & Bolisani, E. (2012). Trust in knowledge exchanges between service providers and clients: A multiple case study of KIBS. *Knowledge Management Research and Practice, 10*(1), 16–26.

Schiuma, G. (2009). The managerial foundations of knowledge assets dynamics. *Knowledge Management Research and Practice, 7*, 290–299.

Senge, P. (1999). *The fifth discipline: The art & practice of the learning organization*. New York: Random House.

Spender, J. C. (1996). Making knowledge the basis of dynamic theory of the firm. *Strategic Management Journal, 17*, 45–62.

Spender, J. C., & Strong, B. A. (2014). *Strategic conversations: Creating and directing the entrepreneurial workforce*. Cambridge: Cambridge University Press.

Stacey, R. D. (2001). *Complex responsive processes in organizations: Learning and knowledge creation*. London: Routledge.

Studdard, N. L., & Munchus, G. (2009). Entrepreneurial firms' acquisition of knowledge using proactive help-seeking behavior. *International Journal of Entrepreneurial Behavior and Research, 15*(3), 242–261.

Sveiby, K. E. (2001). A knowledge-based theory of the firm to guide in strategy formulation. *Journal of Intellectual Capital, 2*(4), 344–358.

Sveiby, K. E. (2010). *Methods for measuring intangible assets*. (Online). Accessed February 10, 2017, from http://www.sveiby.com/articles/intangibleMethods.htm

Szulanski, G. (1995). Unpacking stickiness: An empirical investigation of the barriers to transfer best practice inside the firm. *Academy of Management Proceedings, 1995*, 437–441.

Szulanski, G. (2000). The process of knowledge transfer: A diachronic analysis of stickiness. *Organizational Behavior and Human Decision Processes, 82*(1), 9–27.
Tichy, N. M. (1997). *The leadership engine: How winning companies build leaders at every level*. New York: Harper Business.
Tsai, W. (2001). Knowledge transfer in intraorganizational networks: Effects of network position and absorptive capacity on business unit innovation and performance. *The Academy of Management Journal, 44*(5), 996–1004.
Tsoukas, H. (1996). The firm as a distributed knowledge system: A constructionist approach. *Strategic Management Journal, 17*, 11–25.
Vallejo-Alonso, B., Rodriguez-Castellanos, A., & Arregui-Ayastuy, G. (2011). *Identifying, measuring, and valuing knowledge-based intangible assets: New perspectives*. Hershey, PA: IGI Global.
Van den Berg, C., & Popescu, I. (2005). An experience in knowledge mapping. *Journal of Knowledge Management, 9*(2), 123–128.
Verleye, K. (2015). The co-creation experience from the customer perspective: Its measurement and determinants. *Journal of Service Management, 26*(2), 321–342.
Von Hipel, E. (1994). 'Sticky information' and the locus of problem solving: Implications for innovations. *Management Science, 40*(4), 429–439.
Wenger, E., McDermott, R., & Snyder, W. M. (2002). *A guide to managing knowledge. Cultivating communities of practice*. Boston: Harvard Business School Press.
Yoo, K., Suh, E., & Kim, K. Y. (2007). Knowledge flow-based business process redesign: Applying a knowledge map to redesign a business process. *Journal of Knowledge Management, 11*(3), 104–125.
Zack, M. H. (1999). Developing a knowledge strategy. *California Management Review, 41*(3), 125–145.

8 Strategic Performance and Knowledge Measurement

Measurement has generally considered integral part of strategic management: it is assumed that, to plan the goals of a strategy, we need a method to measure some strategic performances that define the success itself of the strategy, and makes it possible to control its implementation. So, when it comes to knowledge strategies, we may say that we need a way to measure strategic performances related to knowledge. In other words, we need some measurement system that can applied to KM and knowledge. How to measure knowledge has attracted the interest of scholars since more than two decades. Many techniques have been proposed but they are still far from becoming an established practice. They are very heterogeneous, are based on different foundations, and often derive from techniques formerly developed for other goals. A *rationale* to treat the problem still lacks, which is critical for both practice and research. The chapter attempts to give some order to this issue, which is still much debated in the KM literature. Especially, the theoretical and methodological soundness of the various measurement techniques is questioned. By examining the extant literature, we will illustrate the state-of-the-art of knowledge measurement and the related implications for KM research and practice. We will analyze some classifications of the various measurement methods that have been proposed in the literature, and discuss their differences, point of strength and weakness. Also, their soundness from the point of view of a theory of measurement will be examined. Implications for research and practice are summarized in the conclusion.

8.1 Measurement: An Essential Ingredient of Strategies?

The importance ascribed to measurement is not new, and connects thinkers, philosophers, and scientists across millennia. Especially, measurement and measurability have generally been associated to quantities and numbers. Consequently, if numbers represent the "background representation" of reality (a quotation of the Greek philosopher Pythagoras—reported by Iamblichus of Calcis in his "Life of

E. Bolisani, C. Bratianu, *Emergent Knowledge Strategies*, Knowledge Management and Organizational Learning 4, DOI 10.1007/978-3-319-60657-6_8

Pythagoras"—says: "Number is the ruler of forms and ideas, and the cause of gods and daemons"), measures and measurement may represent our ability to understand the world. A famous statement, attributed to the great scientist Galileo Galilei, says: "Measure what is measurable, and make measurable what is not so"; more recently, William Thompson (Lord Kelvin) affirmed that: "I often say that when you can measure what you are speaking about, and express it in numbers, you know something about it; but when you cannot measure it, when you cannot express it in numbers, your knowledge is of a meagre and unsatisfactory kind; it may be the beginning of knowledge, but you have scarcely, in your thoughts, advanced to the stage of science, whatever the matter may be" (Kelvin 1883). To sum up, if we follow these great thinkers and scientists, any effort of measurement would be worthwhile because it may enable us to learn more about the reality around us.

It is easy to understand why measurement is vital in the so-called "hard sciences" like physics. But its centrality is also clear in other human activities. In particular, the importance of measurement for business management and decision making has been openly declared. For example, Hubbard (2007, p. 47) affirms: "If a measurement matters at all, it is because it must have some conceivable effect on decisions and behavior. If we can't identify a decision that could be affected by a proposed measurement and how it could change those decisions, then the measurement simply has no value". Indeed, the literature is full of citations that go to the same direction: for example, Harrington (1987) affirms that "measurement is the first step that leads to control and eventually to improvement. If you can't measure something, you can't understand it. If you can't understand it, you can't control it. If you can't control it, you can't improve it". The diffusion of data processing in companies, and especially the upsurge of Internet applications with new terms that are becoming pretty popular such as "Big Data" and "Business Analytics", are also pushing up the idea that decision-making in modern companies is intrinsically intertwined with the capability to process and analyze data. For some companies, as Davenport (2006a, p. 98) argues, this can be the real reason of their success: "Some companies have built their very business on their ability to collect, analyze, and act on data. Every company can learn from what these firms do". Emblematic is the title of another piece written by Davenport (2006b): "In God we trust: All the others bring data", which appears to be a sort of manifesto for modern management.

Actually, the matter is controversial. Another popular statement "You can't manage what you don't measure" (which is, surprisingly, attributed to different authors—sometimes Edwards Deming, some other times Peter Drucker—see McAfee et al. 2012), is used by different people to draw different conclusions. For example, according to Hunter (2015), while the above mentioned sentence is often used as a strong recommendation to use measures in business decisions, Deming himself apparently warned against an unquestioning use of measures: "Dr. Deming did very much believe in the value of using data to help improve the management of the organization. But he also knew that it wasn't close to enough. There are many things that cannot be measured and still must be managed. And there are many things that cannot be measured and managers must still make

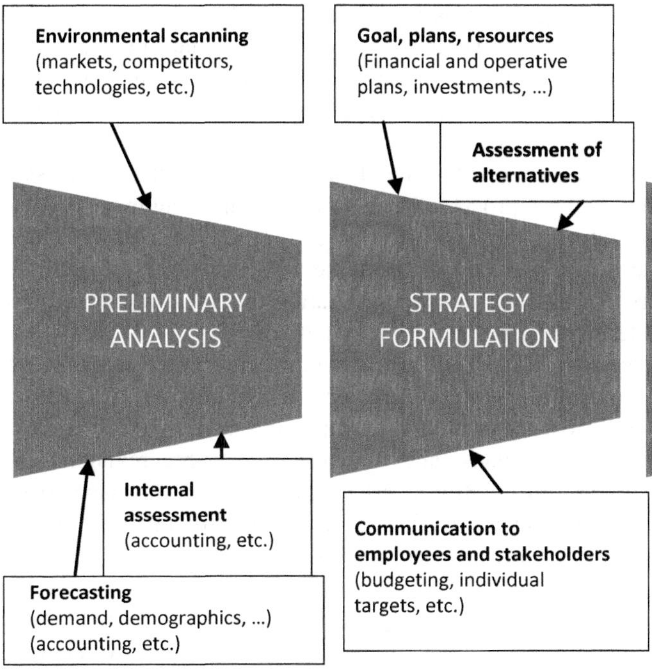

Fig. 8.1 Measurement in strategy formulation

decisions about" (Hunter 2015). As Hunter recalls, his correct and complete citation should be "It is wrong to suppose that if you can't measure it, you can't manage it—a costly myth" (Deming 1994, p. 35).

When it comes to strategies, the point is, obviously, the same. Measurement can be an essential ingredient of strategy formulation (Simons 2013). In the classic "rational" approach, its role is essential in at least three different moments, which however call for different methods and applications (Fig. 8.1).

- In the preliminary analysis: the formulation of a strategy starts from a collection and analysis of information about the "state" of the environment and the organization, and a formulation of plausible forecasts of relevant phenomena. Here, several activities can imply measures of various kinds. For example, in environmental scanning (Morrison 1992), several measurable indicators can be needed to describe markets, competitive context, and other aspects of the social or economic arena: demand, market shares of competitors, demographic indexes, etc. Industrial and financial accounting indexes, and other internal measures describing the state of the organization and its various components, are also essential ingredients. Finally, forecasting techniques often include quantitative extrapolations of indexes that predict future measures of relevant external or internal aspects.

- In the formulation of strategic alternatives: the goals of a strategy are often formulated in quantitative terms which refer to measures that are relevant to stakeholders and executives (profits, costs, market shares, etc.). Similarly, plans and resources include measures of investments, tangible or intangible resources to acquire or develop, etc. Finally, it is worth recalling that financial planning and evaluation are still an important element of strategy formulation (Hussey 1998). As is well known, these processes heavily rest on financial measures and other indicators of investment appraisal. In strategy formulation, measures have also the fundamental function to communicate key objectives and priorities to external stakeholders and to the internal organization that must implement the strategy in practical terms (Micheli and Manzoni 2010).
- In monitoring and control of strategy implementation: the same (or additional) measures, that are employed to define goals, plans, and resources, are also used to assess the "success" of a strategy, for example in terms of the degree of satisfaction of the budgeting plans, and to consider possible revisions and adjustments. Here, strategy formulation overlaps the field of performance measurement in business (Neely et al. 2005), under the assumption that positive business performances are an indicator of the success of a strategy.

In this process, the advancements of technology also have a role. As Micheli and Manzoni (2010) argue "thanks to data collection automation and improvements in data analysis, organizations have introduced increasing numbers of performance indicators". The emergent phenomenon of the so-called "big data" brings about a new faith on our capability to drive decisions, based on a rational and systematic measurement of a plurality of data that, today, are available to companies by means of internal information systems, the Internet and the Web (Provost and Fawcett 2013). In principle, these are big opportunities because "the inclusion of more performance indicators could be welcome" especially by executives that are aware of the complexity of decisions and the dynamism of the environment, which makes it difficult to rely on mono-dimensional measurements that may be unable to synthesize a multi-faceted reality. So, a plurality of indicators, measurements and measurement systems has progressively been introduced in the managerial literature (Franco-Santos et al. 2007; Neely et al. 2005).

On the other hand, too many indicators may lead to difficult and even contrasting or inconsistent interpretations of what we are going to do, what it is going to be, or what is going on in the business. In the end, being too stuck to a huge set of indicators can and kept under control can be detrimental for adaptability and capability to change (Micheli and Manzoni 2010). Some authors (e.g. Eisenhardt and Sull 2001) suggest that, especially in dynamic environments and/or where there is high uncertainty, only a few indicators should be introduced. Indeed, an exceeding number of measures may give the illusion of complete knowledge, because these tend to become uncritically accepted with no discussion of their reliability.

To overcome this difficulty, others have attempted to create "synthetic" indicators by means of a multi-level structure of measures, with the purpose to combine a higher variety of measurement dimensions with a general view and a consistent interpretation. A good example is the balanced scorecard (Kaplan and Norton 1996) that has many followers, but also criticisms. Voelpel et al. (2006) underline several possible drawbacks. The most critical one is that a static and mechanistic view of companies and business would be implied by the scorecard analysis, view that may result in "difficulties of managing certain aspects of corporate life, such as promoting dynamic innovation and knowledge creation". And in addition "In a knowledge driven company, simple cause-effect relationships are not sufficient anymore to understand complex relationships that the BSC tries to reduce to a linear one-way relationship" (Voelpel et al. 2006, p. 54).

The complexity of systems and phenomena under observation is also cause of the proliferation of *qualitative measures* that are used as substitutes or additions to quantitative indicators, especially for special purposes (e.g., measurement of employee-related performances—Morris et al. 1991) or in specific contexts (for example, small firms—Jarvis et al. 2000). But clearly, introducing qualitative measurements has implications both in terms of compatibility and consistency with quantitative indexes, and in relation to the intrinsic subjectivity that may affect the measures themselves (Ittner et al. 2003).

The discussion about the issue of measurement in strategy formulation also implies a reflection on the possible "extreme approaches" to strategy that we analyzed in Chap. 6. A purely rational/deliberate approach implies, *per se*, an ex-ante definition of measurements and measures that are appropriate to a) analyze the context, b) formulate goals and plans, and c) control proper implementation. In other words, not only a pre-defined model of the company in its competitive environment is adopted and used, but also the way of *measuring* this model in its relevant traits. This is challenged by the emergent view of strategies (Lowe and Jones 2004): when we accept that strategies are the ex-post rationalization of successful practices in a company, this also means that the way this business success is measured can't be pre-defined but it also *emerges* from the ground experience of employees. Again, as we argued in Chap. 6, it is likely that relevant business measurements partly come from pre-defined models, and partly emerge from the daily practice and the learned experience. But in any case, integrating and finding a balance between these opposing views represent a real problem for companies.

8.2 Knowledge Measurement for Knowledge Strategies

Considering the main topic of this book, it is necessary to understand if and how the inclusion of knowledge strategies in strategy formulation has implications for measurement. The particular point we will discuss here is *knowledge measurement*, namely, measuring knowledge, its manifestations, and the processes that are used to manage it. The importance of measurement in knowledge strategy formulation can be analyzed by replicating the same arguments that we developed in the previous section. Generally speaking, in knowledge strategies, measurement can be important:

- In *preliminary analysis*, i.e. assessing the "knowledge state" of a company (what and how much we know, what cognitive resources we have in our company, etc.)
- In *strategy formulation*, i.e. setting goals, plans, and resources for KM, learning, and other cognition-related activities
- In *monitoring and control*, i.e. to check how much the knowledge strategy has achieved its goals (i.e. how much we have learnt, how effectively we are managing our knowledge resources, etc.)

However, there is an additional point to consider: if a knowledge strategy is strictly associated with a company strategy, measurement becomes important not only for the knowledge strategy itself, but also for the implications that a knowledge strategy can have in the company strategy. For example, KM processes are not important *per se*, but also because they can improve the quality of work in a company and, therefore, increase some performances of the business. Similarly, the measure of the "amount of knowledge" that we acquire, by means of a learning activity, can enable better business performances. Consequently, knowledge measurement can be related to other measures of performance, in an intertwined system of measurement. An example of this is presented in Fig. 8.2.

8.3 Knowledge Measurement in the Literature

In the previous section, we have argued that, regardless our viewpoint on knowledge strategies, their connection with company strategies, and the approaches to strategy formulation, if we accept the notion of knowledge strategy we must consider the problem of knowledge measurement. In this section, we will focus on the conceptual point that lays in the background: what is really knowledge measurement? What does this mean in practical terms? Is there a sufficiently reliable "method" that can be used for measuring knowledge?

In the knowledge management and intellectual capital (IC) literature, how to measure knowledge has attracted the interest of scholars since at least two decades (Ragab and Arisha 2013). As a result, we have an impressively well equipped toolbox of methods (Liebowitz and Suen 2000; Grossman 2006; Sveiby 2010). Many techniques have been proposed and applied, both by scholars and practitioners. Unfortunately, they are still far from becoming an established practice. They are very heterogeneous, and often derive from techniques formerly developed for other goals.

Even the positions of scholars and practitioners range from optimistic opinions to negative viewpoints about the usefulness and practicability of knowledge measurement. In this puzzling context, a *rationale* to treat the problem still lacks, which is a critical point for both the practice and the research. For the practice, reflecting on the actual applicability of measurement methods to concrete situations is essential: this can enable consistent interpretation, use, and comparisons. Particularly, measuring knowledge can be vital for companies implementing KM and IC

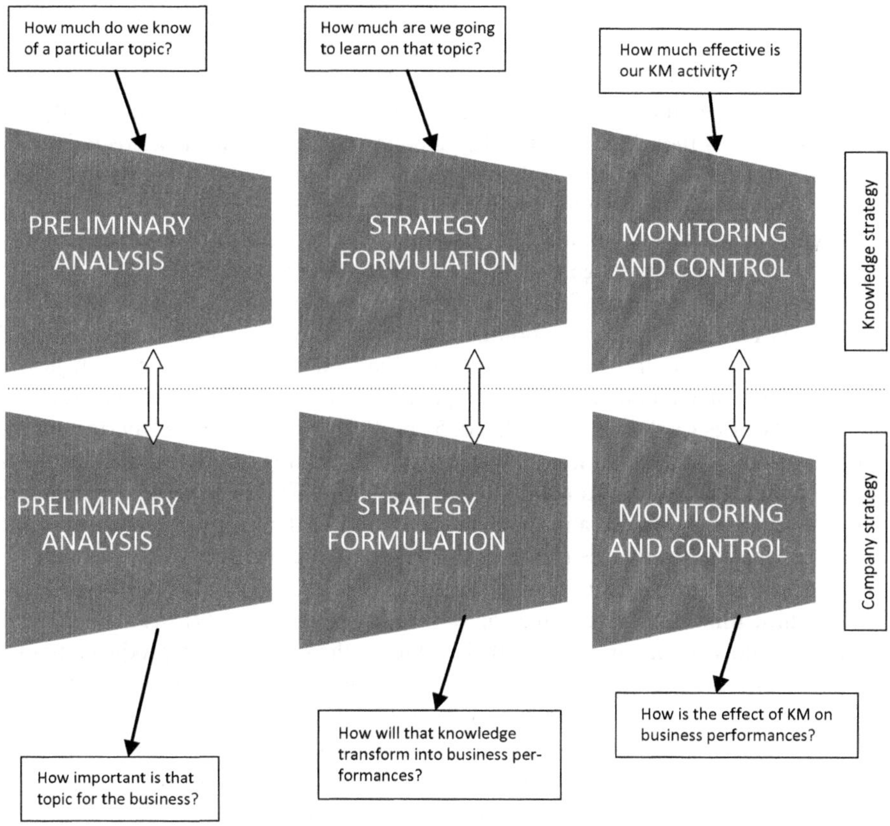

Fig. 8.2 Intertwined measurement system

initiatives. For research, the development of a solid approach to knowledge measurement appears vital for improving theory building and empirical studies.

Similarly to what we argued when we defined the notion of knowledge strategy, there are at least three different fields to which knowledge measurement is relevant (Bolisani and Oltramari 2012). The first field is that of IC studies, which focus on intangibles that, being key constituents of business, require measurement (Hand and Lev 2003). Some "traditional" intangibles (like brands and licenses) have now an established place in business, and are also included in the accounting systems. The literature has later identified that other important category of intangibles that is Intellectual Capital. While this term has become quite popular, its definition has been debated a lot: "The term 'intellectual capital' is frequently used in an all-encompassing fashion with the risk that in time the identity of the object will become unclear" (Petty and Guthrie 2000, p. 158). Therefore, the notion is often considered by splitting it into its main elements (Roos et al. 1997), i.e. human capital (skills, experience, competencies of members of the organization), relational capital (relations that the company and its member have with external entities—customers, suppliers, etc.) and structural capital (processes, information systems, databases, management policies, etc.). What is important from our viewpoint is that IC has, indeed, knowledge as their core ingredient (Stewart 2003). Hence, measuring intellectual capital becomes a matter of measuring knowledge, in some form or another. Particularly, "IC Accounting" represents methods to include IC into financial (or non financial) reports and, according to Guthrie et al. (2012), has a "strong and continuing tradition". So, since the notion of IC is strictly connected with knowledge and KM strategies, attempting to measure IC implies trying to measure knowledge in some form (Petty and Guthrie 2000).

A second field is that of KM programs. The increasing diffusion of KM in companies raises an important point: "there is a need for metrics to justify KM initiatives. Also, linking KM initiatives to financial investment may help justify KM to senior management and thus improve the firm's ability to manage knowledge assets effectively" (Lee et al. 2005, p. 470). Measuring KM performances serves various goals including: budgeting for KM, setting targets, giving feedback for implementation, etc. (Kankanhalli and Tan 2004). This requires a measurement of KM processes (Goldoni and Oliveira 2006) and of their impact on business and organizational performances (Zack et al. 2009; Dalkir et al. 2007). All this calls for specific methods and metrics (Lee et al. 2005; Siong Choy et al. 2006).

The third important field is that of *knowledge-intensive industries*. As we have mentioned, these companies found their value on the capability to produce, manipulate, and provide services incorporating knowledge (Bolisani et al. 2016), such as: IT services, consulting, information, etc. And today even the supply of physical goods often implies the provision of "knowledge contents" that are attached to them: training services, assistance, and the like. Indeed, traditional manufacturing sectors have been long subjected to a progressive transformation, where the distinguishing added value is increasingly based on the knowledge provided to customers in the forms of services. Hence, measuring the value of all these knowledge-intensive services, products or activities that are sold to customers requires measuring knowledge. A critical questions is, for example (Bolisani et al. 2016): what is the value of knowledge incorporated in services provided to customers?

In the literature, knowledge measurement has a special place (Dalkir et al. 2007; Ragab and Arisha 2013; Martin-Decastro 2014). In a recent systematic review of the literature (Bolisani 2016), many authoritative KM and IC journals were examined. Although not all the bibliographic sources were covered, the analysis well illustrates the current state of the research on this topic. Figure 8.3 reports the distribution of pertinent papers across journals, and shows that, based on the annual trend, the issue of knowledge measurement is still central in the KM/IC literature, and the number of papers is still growing.

Another interesting result of the analysis is the classification of the papers (Table 8.1), based on two dimensions. The first one is the kind of study, that can be: *empirical* (where the main focus is on the application of a particular method), *conceptual/theoretical* (where a new measurement method is proposed, old methods are analyzed, or their foundations are discussed), and *review* (analysis of

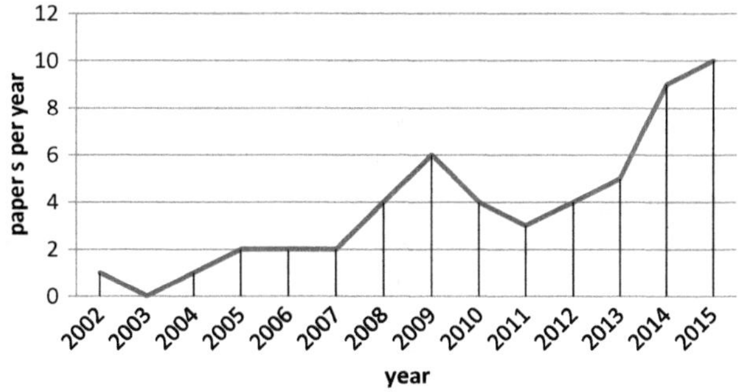

Fig. 8.3 Articles on knowledge measurement in KM and IC journals: annual trend (source: Bolisani 2016)

Table 8.1 Classification of studies of knowledge measurement

Topic	Empirical studies	Conceptual or theoretical studies	Reviews	Total
LEARNING	3	2	0	5
KM	4	8	4	16
IC	12	15	5	32
Total	19	25	9	53

Source: Bolisani (2016)

secondary sources—e.g. literature, existing cases, websites, etc.). The second dimension is the prevailing area, that can be: *learning* (i.e. measurement of learning processes or learning organizations), *KM* (namely, measuring KM performances and processes), and *IC* (that is, IC measurement).

First of all, it is worth to notice that studies of IC measurement prevail, which can be explained considering the long tradition in IC measurement and its strict relationship with accounting practices (Guthrie et al. 2012). This is also important because the transfer of a measurement method from the IC field to others that relate to knowledge can't be taken for granted. For example, it is likely that measurement be more easily applicable to IC, even in quantitative terms, rather than learning (whose definition is, by nature, more ambiguous), or to KM itself, that includes several different aspects and elements, some well-defined others less. Second, conceptual/theoretical studies still prevail, and although it may be difficult to draw ultimate conclusions on the reason for this, it appears a sign that the field of measurement applied to knowledge-related concepts is still looking for conceptual formalization of notions, methods, and approaches.

8.4 Existing Measurement Methods

Researchers and practitioners have produced a broad variety of approaches and techniques (for a reference, see e.g. Malhotra 2003; Boudreau 2003; Grossman 2006; Kuah and Wong 2011; Sveiby 2010). Table 8.2 proposes an outline of some of these methods. Again, the analysis is not complete (there are many more methods

Table 8.2 Methods of knowledge measurement

Name	Proposers and source (original references are in Sveiby 2010)	Type	Details
ICU report	Sanchez (2009)	SC	EU-funded project for universities
EVVICAE	McCutcheon (2008)	DIC	Web-based toolkit
RICI	Schiuma et al. (2008)	SC	Set of indicators for regions
IabM	Japan Ministry of Economic, Trade and Industry (2004)	SC	Indicators based on the MERITUM guidelines
SICAP	(various—2004)	SC	EU funded project for public services
National Intellectual Capital Index	Bontis (2004)	SC	Modified version of the Skandia Navigator
Topplinjen/ Business IQ	Sandvik (2004)	SC	Developed by a consulting company
Public sector IC	Bossi (2003)	SC	Developed for public sector
Danish guidelines	Mouritsen et al. (2003)	SC	Guidelines for Danish companies
IC-dVAL	Bonfour (2003)	SC	Based on four competitive indicators
Intellectus model	Sanchez-Camizares (2007)	SC	
Knowledge Asset Methodology	World Bank (2002)	SC	Measurement of "knowledge-related" assets and enabling factors of countries
FiMIAM	Rodov and Leiaert (2002)	DIC MCM	Monetary value of intangible assets
IC Rating	Edvisson (2002)	SC	Extension of Skandia Navigator
Value Chain Scoreboard	Lev (2002)	SC	Matrix of non-financial indicators
Meritum	Meritum Guidelines (2002)	SC	EU project for disclosure of IC
Intangible asset statement	Garcia (2001)	SC	Measurement of growth/ renovation; targeted to the public sector
Knowledge audit cycle	Schiuma and Marr (2001)	SC	Measures of organizational capabilities
Value creation index	Baum et al. (2000)	SC	Non financial metrics for IC
Value Explorer	Andriessen and Tiessen (2000)	DIC	Accounting methodology of KMPG
Intellectual asset	Sullivan (2000)	DIC	Value of intellectual property
Total value creation	Anderson and McLean (2000)	DIC	Discounted projected cash flow
Knowledge Capital Earnings	Lev (1999)	ROA	Earnings beyond book assets

(continued)

Table 8.2 (continued)

Name	Proposers and source (original references are in Sveiby 2010)	Type	Details
Inclusive Valuation methodology	McPherson (1998)	DIC	Hierarchies of weighted indicators
Accounting for the future	Nash (1998)	DIC	Projected discounted cash flow
Investor assigned market value	Standfield (1998)	MCM	Identification of IC value as a component of a company's stock value
Calculated intangible value	Stewart (1997)	MCM	Stock market value minus book value
Economic Value Added	Stern and Stewart (1997)	ROA	Profit adjusted to intangibles
Value Added Intellectual Coefficient	Pulic (1997)	ROA	Equation estimating value creation from intellectual capital
IC-Index	Roos et al. (1997)	SC	Multiple scorecard indexes
Technology broker	Brooking (1996)	DIC	Diagnosis based on 20 questions
Citation-weighted patents	Dow Chemical (1996)	DIC	Impact of R&D on patent production
Holistic accounts	Ramball Group (1995)	SC	Euro Foundation Quality Management Business Excellence Model
Skandia Navigator	Edvisson and Malone (1997)	SC	164 metrics grouped into 5 categories
Intangible Asset Monitor	Sveiby (1997)	SC	Connection of IC to strategic objectives
HR statement	Ahonen (1998)	DIC	Accounting principles for human capital
Invisible balance sheet	Sveiby (1989) and others	MCM	Stock market value minus book value
HR Costing/ Accounting	Johansson (1996)	DIC	Hidden costs of human resources
Tobin's q	Tobin (1950 and around)	MCM	Stock value per replacement cost of assets

Adapted from Sveiby (2010) and Bolisani (2016)

that are available in the literature) but is sufficient to highlight their extreme heterogeneity, in consideration of their approach, application and general purpose. Sveiby (2010) for example classifies these methods into four different categories:

- Direct Intellectual Capital methods (DIC): these estimate the monetary value of knowledge-related intangibles and their various components. They generally focus on one or more manifestations of company knowledge and IC and measure their economic value.
- Market Capitalization Methods (MCM): they calculate the difference between a company's market capitalization and its stockholders' equity. The general assumption here is that this difference depends on the value of knowledge possessed by the company: the "more valuable knowledge" a company has, the more the investors will recognize this in terms of growth potential, which will reflect on the value of the company in the financial markets.
- Return on Assets methods (ROA): these methods are based on similar principles of the analogous techniques used in financial analysis. The returns of the amount of knowledge and IC possessed by the company (or the amount of investments in knowledge) are measured
- Scorecard Methods (SC): these methods replicate the approaches of Kaplan and Norton's (1996) Balanced Scorecard. The various components of IC and knowledge-related assets or activities are singled out and then measured by using specific indicators or composite indexes, sometimes collected in synthetic tables.

As can be easily understood, being all these methods different in nature, application, meaning, and calculation, they can have peculiar advantages and drawbacks (Sveiby 2010). For example, ROA and MCM methods are useful for stock market valuations or for comparing companies. However, they are substantially based on the assumption that everything can be measured in monetary terms (i.e. costs, prices, market values, etc.), which can be a superficial approach that can neglect the specific nature of knowledge in organizations. DIS and SC methods provide a broader picture of an organisation's health, and since they do not need to measure financial elements necessarily, they can also be applied to non-profit organisations, public services, and even entire nations or regions.

However, DIS and SC indicators are often contextual, and so they must be customised for each organisation or goal, and provide measures that may not be comparable between a situation to another. Also, they may not be easily accepted by companies and managers who are accustomed to pure financial perspectives. Finally, they can generate large amounts of data which are hard to analyse and communicate.

There are also other possible classifications that highlight more differences between the distinct knowledge measurement methods. For example, by adopting a managerial view, Ragab and Arisha (2013) distinguish between methods that adopt an *internal* rather than an *external* perspective: the former have the goal of measuring knowledge to help managerial control, while the latter aims to show the

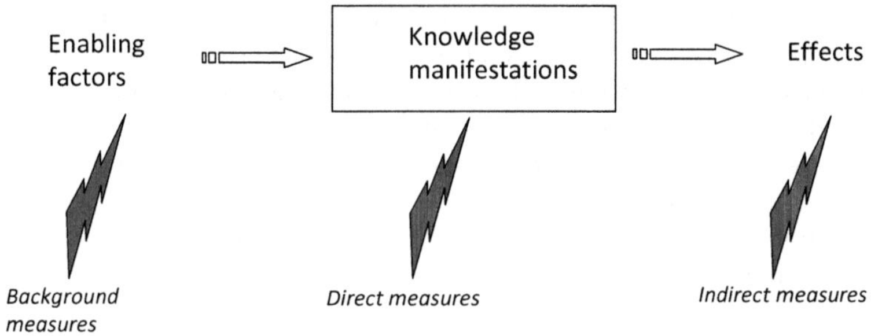

Fig. 8.4 Knowledge measures based on the stage of knowledge production and usage (adapted from Gambarotto et al. 2011; Bolisani 2016)

value of a company's knowledge to external stakeholders. Also, they classify techniques into four categories, namely: *financial models* that calculate the value of knowledge and IC based on financial statements; *IC methods* that identify, classify and evaluate the various IC components separately and with different techniques; *human capital methods* that focus solely on individuals by assuming them as the main components of IC; *performance methods*, which focus on KM processes and their outcomes.

There are also differences regarding the *stages* in the production and exploitation of knowledge. It can first be considered (Fig. 8.4) that knowledge is produced based on some *inputs or enabling factors*: for example, human capital, investments in databases and IT systems, etc. Consequently, it can be assumed that the more companies invest in enabling factor, the more likely is that they can produce valuable knowledge, that may finally transform into better business or organizational performances. For this reason, it is possible to find some measurement methods that substantially use what can be called *background measures* of the *enabling factors* that can favor knowledge production and exploitation: for example, number of qualified employees, investments in IT resources, investments in R&D, etc. In this category it is possible to include popular methods for evaluating knowledge capital of nations and economic systems.

Second, the use of some enabling factors can lead to the production and accumulation of knowledge, that appears in some form or another: for example, projects, documents, patents, etc. So, there are measurement methods that adopt what can be called *direct measures*, that focus on these manifestations of knowledge. Example include quantitative measures of what are often called "knowledge items" or "objects" (Bolisani and Oltramari 2012), like for instance: number of documents produced, patents, number of projects, number of clients in a database, etc. The assumption is that knowledge can often emerge in the form of tangible artifacts which are, consequently, directly assessed.

Third, the knowledge possessed and processes can have some final *effects* that can be measured. Therefore, there are measurement methods that use *indirect measures* of the supposed effects of knowledge, generally on some kind of performance. For instance, it is assumed that the possession of knowledge can influence operational performances: measuring these performances is considered a way to measure the knowledge that cause them. "Indirect" estimates include for example the EVA (Economic Value Added) approach than considers the value of a company's knowledge as its *market over-evaluation* compared to its actual "hard assets" (Grossman 2006).

But a more detailed analysis of existing methods also reveals other specific distinctions. For example:

- *Quantitative vs. qualitative indicators:* There are methods that use quantitative indicators (for example: number or book value of patents, number of graduated employees, investments in IT, etc.) and others that are purely based on qualitative judgments (i.e.: assessments of competences, potential importance of R&D projects, etc.).
- *Focus*: some methods focus on individuals (i.e. number of skilled people, their qualifications, etc.), others on artifacts (such as: patents and property rights, number of projects, documents, IT facilities, etc.), or on processes (for example: effectiveness of document retrieval, training activities, R&D projects, etc.).
- *Unit of measurement:* generally, qualitative methods don't have units of measurement (with the exception of "judgments" that can expressed by means of scores). In quantitative methods, the units of measure can be of various kind: value (money), quantities (number of people or patents, size of a database, and others) even times (e.g. time spent for searching for a document, for participating actively in a community of practice, etc.).
- *Level of analysis:* some approaches consider individuals, others business units, offices, processes, or projects; or it can be the company, the region, the nation.
- *Sources of data:* a first essential source is the internal accounting system, and especially financial statements, that are sources of monetary measurements (e.g.: investments in knowledge-based activities, book value of patents, etc.). Other sources are internal company archives, for example the Human Resource Management department (e.g. profiles on employees, qualifications, etc.), the Sales department (databases of clients), the R&D department (project documentations, etc.), the IT department (details about IT systems), etc. In some cases, special collections of data are arranged with the specific purpose of measuring IC and knowledge-based activities: this is, for instance, typical of KM offices, that can arrange for example questionnaires about the results achieved by people in communities of practice, KM-based processes, etc.
- *Form of indicators:* some indexes are simply numerical, others are combination of numerical and sometimes qualitative indicators.

All these classifications show that, despite the numerous attempts, a convincing and uniform approach to measuring knowledge and its related assets or activities has yet to be developed. This leaves "a gap in the literature that needs to be filled with a clear technique" (Ragab and Arisha 2013, p. 889). The analysis of literature shows that there is awareness of that, but the problem still doesn't have a solution.

Indeed, there are extreme positions about even the feasibility or usefulness of knowledge measurement. On one extreme, there are those who think that measuring knowledge, although challenging, is an achievable goal. According to Hubbard (2007), a champion of this optimistic view, the ultimate purpose of measurement is to contribute to reducing uncertainty: "Measurement [is] a quantitatively expressed reduction of uncertainty based on one or more observations" (p. 23). He adds that this is especially important in decision making: "Why do we care about measurements at all? There are just three reasons. The first reason—and the focus of this book—is that we should care about a measurement because it informs key decisions. Second, a measurement might also be taken because it has its own market value (e.g., results of a consumer survey) and could be sold to other parties for a profit. Third, perhaps a measurement is simply meant to entertain or satisfy a curiosity (e.g., academic research about the evolution of clay pottery). But the methods we discuss in this decision-focused approach to measurement should be useful on those occasions, too. If a measurement is not informing your decisions, it could still be informing the decisions of others who are willing to pay for the information.". Hubbard is aware that any measurement has limitations, but he substantially affirms that it is better than nothing: "Essentially, all models are wrong, but some are useful". Particularly, he contrasts any pessimistic position about the possibility to measure intangibles—including, *ça va sans dir*, IC and knowledge: "The word "intangible" has also come to mean utterly immeasurable in any way at all, directly or indirectly" (p. 3) and he adds that "Intangibles that appear to be completely intractable can be measured".

The opposite position have those that consider knowledge measurement not only useless, but even dangerous. They underline that the lack of a rigorous and shared method can be detrimental or misleading (and especially in business) because measurements are affected by intrinsic subjectivity, unavoidable ambiguity, and irreducible uncertainty (Lambe 2004; Gowthorpe 2009). Particularly, by examining the causes of the bankruptcy of the US energy corporation *Enron*, Lambe (2004) warns about the risk to apply accounting measures to IC and other intangibles, which, according to the author, can leads to misinterpretation of data and even to voluntary distortion of markets. This author uses ultimate words about the prospective to include measurements of IC and knowledge into classic accounts: "the task of the accounting profession is not how to learn to count intangible assets, in the mistaken belief that once petrified, they will behave in tamer and more predictable ways. That outcome is unlikely, and to pretend that risk can be diminished merely by quantifying it is foolish and mistaken" (Lambe 2004, p.10).

Intermediate opinions have those (e.g. Stone and Warsono 2003; Skinner 2008) that confirm some doubts but also think that the adoption of some measurement methods can, at least, complement the traditional performance measurement and accounting system that are adopted in the business context. Skinner (2008), for example, warns that the idea of changing the accounting models to include measures of intangible, IC and other knowledge-related assets may be risky: he argues (p. 202) that "the proposals to mandate additional disclosure in the intangibles area are likely to be unsuccessful because of the fact that the nature and measurement of intangibles varies considerably across industries as well as for other reasons"; in addition, expanding the "existing asset recognition criteria to include intangibles currently excluded from balance sheets" can be problematic in a number of respects. Nonetheless, he also underlines that "there are market-based incentives for companies to voluntary provide" a disclosure of information regarding their intangibles: therefore, it may be concluded that there are good reasons for companies at least to experiment methods of measuring IC and other intangibles, maybe on a voluntary basis and not as a pure replacement of other more established and recognized methods.

8.5 Knowledge Measurement and Theories of Measurement

The great efforts made to define methods of knowledge measurement are an element of richness: considering that the cases and situations of companies can be different—as well as the kinds of intangibles, knowledge-related assets and activities—the availability of a great number of tools and techniques can make it possible to choose the one that fits the distinct needs of measurement and application. On the other hand, all this also reflects the complexity of the issue, where a universally applicable method is hard to discover.

Is this a sign of "immaturity", in terms of theoretical and methodological foundations, of the KM field? To address this question, it is useful to re-frame the problem of knowledge measurement into the broader picture of a "theory of measurement". What will we measure when we talk about knowledge measurement? What definitions, concepts, and methods should be applied? A detailed analysis is beyond the scope of this book, but a brief outline of some remarkable points can help to understand the essence of the problem.

As is well known, the issue of measurement is not recent, but the perspective on what can be measured, how, and for what purpose has progressively changed (Diez 1997a, b; Filkenstein 2003). After the initial contributions, centuries ago, of ancient astronomers and architects—and, much later, of the "fathers" of modern science (*in primis* Galilei and Newton), the research on measurement became a specific field of analysis. In the nineteenth century, the axiomatic foundations of a theory of measurement were developed. In addition, pushed by the rapid advancements in theoretical science but also in applied technologies, a rich toolbox of measurement techniques, devices and theoretical conceptualisations appeared in practically any field of human knowledge—not only in the classic areas of Physics or Astronomy, but also in Social sciences, Economics, Psychology, etc.

In theoretical terms, those who study measurement must consider questions such as: what is measurement? Is it possible to have an abstract and universal notion of measurement that is applicable to the various different fields? A definition of measurement that well summarizes and represents the efforts in this area is proposed in Diez (1997a): measurement can be seen as a process by which values are assigned to "objects", "events" or "phenomena", for representing properties that can be referred to as "*magnitudes*" or "*quantities*". If this definition appears clear and unambiguous, it is also evident that applying this concept to the so many different situations can be complex. For example, where we have a "tangible" idea of many objects or events in Physics, the same can't be said in Psychology or Sociology. Therefore, it may be necessary to specify the meaning of measurement and its possible application in the different fields of human knowledge. Filkenstein (2003) analyses a distinction between strongly and weakly defined measurement. A "strongly defined measurement" (Finkelstein 2003) is one whose application is, primarily, to the so called "hard sciences" (e.g. physics, astronomy, chemistry, material sciences, engineering). Strongly defined measurement is characterized by: "(i) precisely defined empirical operations, (ii) mapping on the real number line on which an operation of addition is defined, (iii) well-formed theories for broad domains of knowledge". Measurement is essential for the empirical validation of theories, because it allows to verify laws that can explain phenomena. Under this perspective, we should speak of measurement only when the numbers assigned to objects or qualities adequately represent empirically verifiable relationships.

The application of a strongly defined measurement systems has, therefore, precise requirements that can't be easily satisfied. However, starting from the twentieth century, the research in "new" scientific fields—and particularly the "social sciences" (i.e.: economics, sociology, psychology) has highlighted a great number of contexts and phenomena to which it would be important to apply some kind of measurement. As Finkelstein notes (2009; p. 1271), even though "the descriptive and explanatory power of the physical sciences made them a model for endeavours to extend the same concepts and methods to psychological and social domains", on the other hand "the classic view of measurement was inadequate for the purpose and a wider concept of measurement was developed". So, important efforts were made to extend the notion of measurement to fields where it is hardly possible to meet the conditions for "strongly defined measurement".

So, a broader and "weaker" notion of measurement can be proposed. This notion becomes more applicable to the "soft systems" of social sciences (Filkenstein 2003) contrasted to the "hard systems" of physical (and related) sciences. In a *weak meaning*, measurement is the "*descriptive* representation of the attributes of objects and events of the real world by symbols on the basis of an objective empirical process" (Filkenstein 2009). The strict correlation between numbers representing objects and the empirical laws connecting them, as typically required in hard sciences, is relaxed. Additionally, weak measurement is based on an "ill-defined concept of the quality to be measured", and there is acceptance of a reasonable amount of uncertainty in the system of empirical relations that it represents. A comparison between the main traits of strongly and weakly defined measurement is proposed in Table 8.3.

When it comes to knowledge measurement (but also, more broadly, to the measurement of related concepts such as IC and many other forms of intangibles that have significance in business), it is clear that a "strong" approach is hard to apply. First, there is no agreed formal theory of what knowledge is and what attributes should or can be measured. Second, and as a consequence, the possibility

Table 8.3 Strongly vs. weakly defined measurement

Strongly defined measurement	Weakly defined measurement
Precisely defined empirical operations Mapping on the real number line on which an operation of addition is defined Well-formed and complete theories for broad domains of knowledge The definition of the quantities and the empirical relational system are based on the theories The symbolic relational system is rich	Based on an ill-defined concept of the quality to be measured Significant uncertainty in the empirical relational system that the measurement represents Symbolic relational system has limited relations defined on it No adequate theory relating the measurement to other measurements in the same (or other) domain(s)

Source: Filkenstein (2003, 2009)

to formulate consistent laws about properties of knowledge and to verify these empirically by using measures—in the same way as is done in physics—is, apparently, not yet at the reach of researchers. We should, however, consider the idea of *weak measurement* as a way out: in weak measurement, we don't need complete and rigorous definitions, theories, qualities of objects or phenomena. Also, a qualitative/descriptive (and not only quantitative) representation of the elements to be measured can be sufficient in most cases: for example, an ordinal measure (for instance, being able to state that an individual "possesses more knowledge" than another will be more than satisfactory in many cases, and we may not need cardinal measures—i.e. saying that an individual has "double the knowledge" of another).

But even if we consider the idea of weak measurement, there is still the necessity that the measurement system meets some *essential conditions* that Filkenstein (2003) calls "the pragmatics of measurement". By developing and adding to Filkenstein's works, these *conditions* can be summarized as follows: *a) clarity; b) objectivity; c) theoretical soundness;* and *d) generality*. It is now possible to consider these conditions and, by analyzing the current state-of-the-art of existing methods for knowledge measurement, to delineate what problems still need to be solved (Bolisani 2016).

Clarity, i.e. "Measurement must be enough free of ambiguity or vagueness, which especially implies that a clear definition is provided of the entities or properties that are subjected to the measuring process". From this viewpoint, the situation in the KM field is confusing. Many of the methods that have been proposed in the literature (see Table 8.2) don't really measure knowledge but, rather, surrogates (for example, enabling factors like number of researchers in a company). Their measurement may be linked to "a measure of knowledge", but it is unclear how. Even in the IC literature, where the issue of measurement has a longer tradition, an agreed and standard definition of IC still doesn't exist (Chang and Hsieh 2011; Khalique et al. 2011). In addition, the place of knowledge in it is vague. Also, the literature is full of alternative terms, such as for instance: knowledge assets, knowledge resources, and many others. All this contribute to the ambiguity of the concept. Indeed, when it comes to "knowledge measurement", an appropriate definition of knowledge is necessary. However "Knowledge is an abstract concept that is difficult to define due to its many meanings and interpretations that depend on the experience of people, their values and cultures, their education level, and mostly on metaphors used to describe it" (Bratianu 2015, p. 8). Many views of knowledge have been proposed; some are more formal, others are expressed only in natural language with its intrinsic vagueness. What's more, many attributes or properties have been proposed to describe the different ways in which knowledge can be represented (Holsapple 2003).

Objectivity, i.e.: "Measures should be independent from the observer, invariant and indisputable in logical discourse. The measurement process should be replicable". Here, a complication is that knowledge can be sometimes seen as an *object*, i.e. the input or result of a cognitive activity and that can be isolated from the people that process them; in other cases, it is seen as a cognitive *process*, i.e. it loses any meaning when it is separated by the people (Iandoli and Zollo 2007). The *knowledge=object* case is apparently the best candidate for an objective measurement approach, because measurement becomes a matter of *counting* a number of items, or *assessing* the magnitude of their qualities, etc. Apparently, this is an ideal situation that makes it possible to use standard measurement units and a repeatable measurement process. The adoption of a *knowledge=object* view is, however, a sort of "shortcut": it consider just the most tangible manifestations of knowledge (i.e.: patents, licences, documents, posts in a forum, etc.) that are measurable or countable (Goldoni and Oliveira 2006). In other words, rather than reflecting on the actual measurability of knowledge, there is an attempt to isolate only what is easier to measure. This has some evident drawbacks. First, the nature of all "tangible" artefacts that can represent a manifestation of knowledge appears very heterogeneous and not uniform. Secondly, many authors argue that the largest portion of knowledge is that embedded in individuals, i.e. its *tacit* component (for example Stenmark 2000). But how can this be measured without introducing substantial subjectivity into measurement? Or should we just measure "the people" that possess knowledge, or their cognitive activities? All this may give a pretty vague and subjective meaning to both the measurement process and its results.

Theoretical soundness, i.e.: "Measurement should be linked to the existence or validation of relationships between qualities/properties". In research, using measures is important for supporting theories about phenomena that are relevant to KM. This means, for example, linking together measures of enabling factors, knowledge manifestations, and effects of their exploitation (see Fig. 8.4). Similarly, in the practice, knowledge measurement is not important *per se* but, rather, when it is linked with its causes and potential effects (particularly, outcomes of an activity). This is essential for decision making: for example, it is important to have verifiable linkages between the measure of some enabling factors in an organisation (e.g.: quantity of skilled people, amount of investments in IT or R&D) and the "quantity" or "quality" of knowledge produced. Similarly, it would be important to associate the knowledge an organisation possesses with its potential effects on organisational performances. To understand the current problems in defining cause-effect relationships that regard knowledge and its use in organisations, it is again useful to analyse the methods proposed in the literature. As mentioned, these tend to measure knowledge, its causal or enabling factors, and its effects separately by means of different approaches and notions. Can these different measures be connected to one another to build consistent theories? Here, many problems arise. Firstly, as mentioned, the methods that focus on directly "tangible manifestations" of knowledge (such as patents, projects, documents, etc.) can underestimate other manifestations (procedures, know how, but also experience, feelings, etc.) that are not tangible and, very often, not identifiable in the same way. In other words, the total "amount of knowledge" that a company or an individual possesses is often a bundle of different elements. Even if we restrict the application of measurement to tangible manifestations, their connection to causal factors or effects is hard to find. For instance, it may be sensible to argue that the number of qualified people can enhance the capability of a company to produce tangible knowledge elements (e.g. patents) and this can lead to some economic performance (for instance, profits). But the quantitative laws connecting their respective measures are not so clear. Similarly, measuring the performances of individuals in a particular situation can be seen as an attempt to measure the knowledge they possess: the assumption is that there is some "law" connecting this knowledge with the performances it enables. But there is still much work to do to discover these laws, supposing that their validity can be proven.

Generality, i.e.: "Measurement should not be too narrow in terms of its applications". This means providing standard measurement techniques for a wide field of application, and not simply for specific situations and narrow cases. Instead, as emerges from the literature, the latter approach is what is often adopted: peculiar methods and techniques are defined and used for solving particular situations and cases. Their application is, therefore, limited to the piece of research for which it has been designed, or for the company where it is supposed to be used for practical purposes (for instance, assessing its internal KM practices). Here, a comparison between measures becomes difficult, as well as their analysis and discussion by other researchers or practitioners. The results of measurement and their use are confined to the case in question, and may little improve the understanding of phenomena and their implications. Indeed, it must be noticed that, in the IC field at least, there has been some effort of generalization of measurement methods, especially as regards IC accounting. However, this effort has sometimes be seen as too ambitious and potentially dangerous, because, as we recalled, it may induce a false idea of uniformity and standardization that instead, according to some authors (e.g. Gowthorpe 2009), is very far from being achieved.

8.6 Perspectives

In this chapter, we have discussed the importance and feasibility of knowledge measurement, which is an important element of the formulation and implementation of a knowledge strategy. In classic strategic management, although there is awareness of the uncertainties that can affect a measurement process, accounting and performance measures are still considered a key ingredient, or at least one of the references that strategists use to define and keep under control the strategic process. Therefore, it is important to consider if the same conclusion can be drawn in the case of knowledge strategies.

We have however shown that this issue is quite debated. Knowledge measurement is a matter that still deserves theoretical and practical analysis. The state-of-the-art of measurement methodologies provides lessons about the prospects of this field of study, and food for thought to both researchers and practitioners. A very basic aspect on which it is important to reflect is the seeking of foundations of knowledge measurement in terms of a theory of measurement. It may be difficult to talk of knowledge measurement as a "strongly defined measurement system" like those used in the hard sciences. A definition of knowledge measurement as a "weakly defined" system still appears possible, but provided that this definition respects the essential requirements of clarity, objectivity, theoretical soundness, and generality. A useful exercise for the proposers of knowledge measurement methods is to clarify how their methods meet these conditions.

The need to define effective measurement methods can be seen to be urgent for the scientific progress of knowledge management. Knowledge may need measuring both as an economic asset itself, and in relationship to the performances of KM processes. But is this an achievable goal? For sure, we should be aware that a complete and well defined measurement system can be hard or impossible to achieve. So, measuring knowledge and its contribution to a strategy in business can be, at least, an effort to build a mental reference that we can use in strategy formulation and control. However, this will be an imperfect process, strongly affected by uncertainties and ambiguities. But if we won't fall in the trap of an uncritical faith in rationality and we will keep our mind open, a genuine effort of measurement can provide, at least, food for though and also help us to understand the limitations of measurement itself, when it is applied to knowledge and KM.

The variety of methods proposed in the literature so far represents an element or richness and an essential starting point for future developments, but the limitations deriving from this heterogeneity are many. In particular, the lack of a standard approach to measurement makes comparisons difficult, and hinders the applications in cases that are different from the original situation for which a method has been designed. With regard to this, a challenge for researchers can be the extensive experimentation of the various methods to different cases and companies, for testing their applicability and meaningfulness.

As a matter of fact, the way this field will develop may take two directions that are strongly influenced not only by the advancements of the conceptual models of KM, but also by the behaviour itself of companies and, especially, the attitude of KM practitioners towards the application of knowledge measurement in their companies. A first possibility is that practitioners involved in KM, being aware of the difficulty to measure knowledge in a standard way, adopt a "best of breed" approach, depending on the specific goals and objects. This "pragmatic" approach is somewhat sensible, but may increase the heterogeneity of the current methods.

A second possibility is that a general standard way to measure knowledge and KM will, in the end, be found. The possibility to achieve these results is strongly based on the advancements in the conceptual modelling of KM activities and processes. Indeed, the foundations of any measurement system rest on robust conceptual representations of the reality that has to be measured. Other managerial branches (from accounting to production) are based on their own formal models, with which the measurement process becomes possible, and meaningful for the current practice. But this is still a significant challenge not only for practitioners but also for KM researchers.

References

Bolisani, E. (2016). Methods of measuring knowledge: Analysis and classification. In *European Conference on Knowledge Management*. Academic Conferences International Limited

Bolisani, E., & Oltramari, A. (2012). Knowledge as a measurable object in business contexts: A stock-and-flow approach. *Knowledge Management Research and Practice, 10*(3), 275–286.

Bolisani, E., Donò, A., & Scarso, E. (2016). Relational marketing in knowledge-intensive business services: An analysis of the computer services sector. *Knowledge Management Research and Practice, 14*(3), 319–328.

Boudreau, J. (2003). Strategic knowledge measurement and management. In S. Jackson, M. A. Hitt, & A. DeNisi (Eds.), *Managing knowledge for sustained competitive advantage*. New York: Wiley.

Bratianu, C. (2015). *Organizational knowledge dynamics: Managing knowledge creation, acquisition, sharing, and transformation*. Hershey, PA: IGI Global.

Chang, W. S., & Hsieh, J. J. (2011). Intellectual capital and value creation – is innovation capital a missing link? *International Journal of Business and Management, 6*(2), 3–12.

Dalkir, K., Wiseman, E., Shulha, M., & McIntyre, S. (2007). An intellectual capital evaluation approach in a government organization. *Management Decision, 45*(9), 1497–1509.

Davenport, T. H. (2006a). Competing on analytics. *Harvard Business Review, 84*(1), 98–107.

Davenport, R. (2006b). In God we trust: All others bring data. *Training and Development (T+D), 4*, 44–50.

Deming, W. E. (1994). *The new economics* (2nd ed.). Boston, MA: MIT Press.

Diez, J. (1997a). A hundred years of numbers. An historical introduction to measurement theory 1887–1990 – part I. *Studies in History and Philosophy of Science, 28*(1), 167–185.

Diez, J. (1997b). A hundred years of numbers. An historical introduction to measurement theory 1887–1990 – part II. *Studies in History and Philosophy of Science, 28*(2), 237–265.

Eisenhardt, K. N., & Sull, D. N. (2001). Strategy as simple rules. *Harvard Business Review, 79*(1), 106–116.

Filkenstein, L. (2003). Widely, strongly and weakly defined measurement. *Measurement, 34*(1), 39–48.

Filkenstein, L. (2009). Widely-defined measurement – an analysis of challenges. *Measurement, 42* (9), 1270–1277.

Franco-Santos, M., Kennerley, M., Micheli, P., Martinez, V., Mason, S., Marr, B., Gray, D., & Neely, A. (2007). Towards a definition of a business performance measurement system. *International Journal of Operations & Production Management, 27*(8), 784–801.

Gambarotto, F., Bolisani, E., & Scarso, E. (2011). Mapping knowledge territories for development policies. *International Journal of Knowledge-Based Development, 2*(1), 16–33.

Goldoni, V., & Oliveira, M. (2006, August 4–6). *Metrics for knowledge management process* (pp. 1720–1728). Proceedings of the 12th Americas Conference on Information Systems, Acapulco.

Gowthorpe, C. (2009). Wider still and wider? A critical discussion of intellectual capital recognition, measurement and control in a boundary theoretical context. *Critical Perspectives on Accounting, 20*(7), 823–834.

Grossman, M. (2006). An overview of knowledge management assessment approaches. *Journal of American Academy of Business, 8*(2), 242–247.

Guthrie, J., Ricceri, F., & Dumay, J. (2012). Reflections and projections: A decade of intellectual capital accounting research. *The British Accounting Review, 44*(2), 68–82.

Hand, J. R. M., & Lev, B. (2003). *Intangible assets: Values, measures, and risks*. Oxford: Oxford University Press.

Harrington, H. J. (1987). *The improvement process: How America's leading companies improve quality*. New York: McGraw-Hill.

Holsapple, C. W. (2003). Knowledge and its attributes. In C. W. Holsapple (Ed.), *Handbook of knowledge management* (pp. 165–188). Berlin: Springer.

Hubbard, D. (2007). *How to measure anything*. Hoboken, NJ: Wiley.

Hunter, J. (2015). *Myth: If you can't measure it, you can't manage it*. The W. Edwards Deming Institute Blog. Online. https://blog.deming.org/2015/08/myth-if-you-cant-measure-it-you-cant-manage-it/

Hussey, D. E. (1998). *Strategic management*. Oxford: Oxford University Press.

Iandoli, L., & Zollo, G. (2007). *Organizational cognition and learning*. Hershey, PA: IGI.

Ittner, C. D., Larcker, D. F., & Meyer, M. W. (2003). Subjectivity and the weighting of performance measures: Evidence from a balanced scorecard. *The Accounting Review, 78*(3), 725–758.

Jarvis, R., Curran, J., Kitching, J., & Lightfoot, G. (2000). The use of quantitative and qualitative criteria in the measurement of performance in small firms. *Journal of Small Business and Enterprise Development, 7*(2), 123–134.

Kankanhalli, A., & Tan, B. C. Y. (2004). A review of metrics for knowledge management systems and knowledge management initiatives. In *Proceedings of the 37th Hawaii International Conference on System Science*. Los Alamitos: IEEE Computer Society Press.

Kaplan, R. S., & Norton, D. P. (1996). *The balanced scorecard: Translating strategy into action*. Boston, MA: Harvard Business Press.

Kelvin, W. T. (1883). *Electrical units of measurement* (Vol. 1). Institution of Civil Engineers Popular Lectures.

Khalique, M., Shaari, N., Abdul, J., & Isa, A. H. B. M. (2011). Intellectual capital and its major components. *International Journal of Current Research, 33*(6), 343–347.

Kuah, C. T., & Wong, K. Y. (2011). Knowledge management performance measurement: A review. *African Journal of Business Management, 5*(15), 6021–6027.

Lambe, P. (2004). *Accounting for knowledge management* [WWW document]. Accessed March 24, 2016, from http://greenchameleon.com/thoughtpieces/account.pdf

Lee, K. C., Lee, S., & Kang, I. W. (2005). KMPI: Measuring knowledge management performance. *Information Management, 42*(3), 469–482.

Liebowitz, J., & Suen, C. Y. (2000). Developing knowledge management metrics for measuring intellectual capital. *Journal of Intellectual Capital, 1*(1), 54–67.

Lowe, A., & Jones, A. (2004). Emergent strategy and the measurement of performance: The formulation of performance indicators at the microlevel. *Organization Studies, 25*(8), 1313–1337.

Malhotra, J. (2003, September 4–5). *Measuring knowledge assets of a nation: Knowledge systems for development*. Invited Keynote Presentation, United Nations Advisory Meeting of the Department of Economic and Social Affairs, New York.

Martín-de Castro, G. (2014). Intellectual capital and the firm: Some remaining questions and prospects. *Knowledge Management Research and Practice, 12*(3), 239–245.

McAfee, A., Brynjolfsson, E., Davenport, T. H., Patil, D. J., & Barton, D. (2012). Big data. The management revolution. *Harvard Business Review, 90*(10), 61–67.

Micheli, P., & Manzoni, J. F. (2010). Strategic performance measurement: Benefits, limitations and paradoxes. *Long Range Planning, 43*(4), 465–476.

Morris, M. H., Davis, D. L., Allen, J. W., Avila, R. A., & Chapman, J. (1991). Assessing the relationships among performance measures, managerial practices, and satisfaction when evaluating the salesforce: A replication and extension. *Journal of Personal Selling & Sales Management, 11*(3), 25–35.

Morrison, J. L. (1992). Environmental scanning. In M. A. Whitely, J. D. Porter, & R. H. Fenske (Eds.), *A primer for new institutional researchers* (pp. 86–99). Tallahassee, FL: Association for Institutional Research.

Neely, A., Gregory, M., & Platts, K. (2005). Performance measurement system design: A literature review and research agenda. *International Journal of Operations & Production Management, 15*(4), 80–116.

Petty, R., & Guthrie, J. (2000). Intellectual capital literature review: Measurement, reporting and management. *Journal of Intellectual Capital, 1*(2), 155–176.

Provost, F., & Fawcett, T. (2013). Data science and its relationship to big data and data-driven decision making. *Big Data, 1*(1), 51–59.

Ragab, M., & Arisha, A. (2013). Knowledge management and measurement: A critical review. *Journal of Knowledge Management, 17*(6), 873–901.

Roos, J., Roos, G., Dragonetti, N., & Edvinsson, L. (1997). *Intellectual capital. Navigating the new business landscape*. London: MacMillan.

Simons, R. (2013). *Performance measurement and control systems for implementing strategy text and cases*. Pearson New International Edition. Pearson Higher Ed.

Siong Choy, C., Kuan Yew, W., & Lin, B. (2006). Criteria for measuring KM performance outcomes in organisations. *Industrial Management & Data Systems, 106*(7), 917–936.

Skinner, D. J. (2008). Accounting for intangibles – a critical review of policy recommendations. *Accounting and Business Research, 38*(3), 191–204.

Stenmark, D. (2000). Leveraging tacit organizational knowledge. *Journal of Management Information Systems, 17*(3), 9–24.

Stewart, T. A. (2003). *The wealth of knowledge: Intellectual capital and the twenty-first century organization*. New York: Currency Doubleday.

Stone, D. N., & Warsono, S. (2003). Does accounting account for knowledge? In C. W. Holsapple (Ed.), *Handbook of knowledge management* (Vol. 1). Berlin: Springer.

Sveiby, K. E. (2010). *Methods for measuring intangible assets* [WWW document]. Accessed March 1, 2016, from http://www.sveiby.com/articles/IntangibleMethods.htm

Voelpel, S. C., Leibold, M., & Eckhoff, R. A. (2006). The tyranny of the Balanced Scorecard in the innovation economy. *Journal of Intellectual Capital, 7*(1), 43–60.

Zack, M., McKeen, J., & Singh, S. (2009). Knowledge management and organizational performance: An exploratory analysis. *Journal of Knowledge Management, 13*(6), 392–409.

Beyond Conclusion

9

Books end usually with some conclusion which emphasize the main contribution of the authors. That is also a standard request for published papers in international journals. It is a consequence of the western culture of breaking down the reality into pieces and events and searching for their proprieties, and more generally, adopting a precise order for the thread of reasoning. Each event has got a beginning and end. It is so embodied into our mind that we take it for granted to be natural. But if we change the perspective and consider the reality surrounding us as an endless whole then we have to consider events in their transformations, and each end as a new beginning. In that perspective, the present concluding chapter of this book is just a new introduction to another possible book. But more important, it is a new beginning of understanding how to think about the future and how to strategize in the knowledge management domain.

Knowledge strategies emerge at the interaction of strategic thinking with knowledge management and constitute a necessity for understanding business complexity in turbulent times and achieving a competitive advantage. The present book is a journey in the new domain of knowledge management and in the efforts managers make to plan for the future and for their business competitiveness. That means to understand the nature of future and the essence of strategizing. Each chapter of the book focuses on a specific topic but from an integrated perspective. Each chapter makes use of exploitation of the available knowledge published in books and scientific journals, as well as of our experience coming from empirical research performed in different companies and public institutions. At the same time, each chapter is an exploration of possible developments in knowledge management and business strategies. And that means a conceptual experiment in the complex field of knowledge management. The topic of the book has been a real challenge for us and for sure it is a challenge for readers. So, this book won't be put into a standard framework because it will not fit. It is much better to challenge our imagination and our understanding about the future.

Knowledge strategies involve time as a fundamental variable and consider the future as a dynamic realization of the present efforts. Future is not a linear

E. Bolisani, C. Bratianu, *Emergent Knowledge Strategies*, Knowledge Management and Organizational Learning 4, DOI 10.1007/978-3-319-60657-6_9

extrapolation of the present or the unfolding of a given destiny, but an evolving complex pattern in a turbulent world. It is an unknown territory of business development but it can be approached if we understand how to integrate our experience with our imagination and strategize to achieve a competitive advantage. Forget about deterministic planning and rigid strategies. Knowledge strategies should be a result of a dynamic integration of deliberate design and emergent response to new contexts and requirements. At the same time, knowledge strategies should be a generic learning process able to offer solutions to future problems and new knowledge for strategists.

The final challenge revealed by this book regards how to manage knowledge even if you cannot measure it. Forget about measuring knowledge by using linear metrics. All the projects proposed so far to measure the complexity of knowledge by using tangible objects as proxy, or the classical accounting system, showed severe limitations, or even ended in failures. We need to design nonlinear metrics and new systems of valuation knowledge, in the same way in which people needed during the history to measure time and invented, for that, the clock.

To sum up, this book reflects our belief that researching in the field of knowledge management is a fascinating, but also risky, enterprise. Fascinating, because there is still much to analyze, to research, and to discover, which is especially attracting for open-minded and curious people. Risky, because definitions and notions are still unstable and, especially, the practical applications of research can suffer from this instability. So, we are aware that we are moving in a still unexplored territory where nothing can't be taken for granted. But, at least, this makes researchers feel like explorers: and that's the real beauty.

MIX
Papier aus verantwortungsvollen Quellen
Paper from responsible sources
FSC® C105338

Printed by Books on Demand, Germany